BRITISH GEOLOGICAL SURVEY
Natural Environment Research Council

British Regional Geology

The Midland Valley of Scotland

THIRD EDITION

By I. B. Cameron, BSc, and
D. Stephenson, BSc, PhD

LONDON HER MAJESTY'S STATIONERY OFFICE 1985

1835 Geological Survey of Great Britain

150 Years of Service to the Nation

1985 British Geological Survey

On 1 January 1984 the Institute of Geological Sciences was renamed the British Geological Survey. It continues to carry out the geological survey of Great Britain and Northern Ireland (the latter as an agency service for the government of Northern Ireland), and of the surrounding continental shelf, as well as its basic research projects. It also undertakes programmes of British technical aid in geology in developing countries as arranged by the Overseas Development Administration.

The British Geological Survey is a component body of the Natural Environment Research Council.

First published 1936
Second edition 1948
Third edition 1985

ISBN 0 11 884365 6

Foreword to the Third Edition

The first edition of this regional guide was written by the late Dr M. Macgregor and Dr A. G. MacGregor and was published in 1936. A second edition, incorporating minor amendments, was published in 1948. Since then the accumulation of new information and the publication of major research on numerous topics concerning the region has necessitated a complete revision of the text. Mr I. B. Cameron wrote Chapters 1, 4, 6–9, 11, 14, 15 and part of 16. Dr D. Stephenson wrote Chapters 2, 5, 12, 13 and the rest of 16, and Dr J. D. Floyd and Dr R. B. Wilson wrote Chapters 3 and 10 respectively. The book was edited by Dr R. B. Wilson.

The authors wish to record their gratitude for the assistance and advice given by their colleagues in the Survey during the preparation of this guide.

Our thanks are due to the following copyright holders for permission to reproduce and draw upon illustrative material: Cambridge University Press for Figures 8, 16 and 31, which are based on maps in Smith, Hurley and Briden, 1981; The Geologists' Association for Figure 41, which is taken from fig. 1 in Dawson, 1980; The Scottish Academic Press Ltd for Figure 4 (fig. 2 in Bamford *in* Harris, Holland and Leake, 1979); The Scottish Journal of Geology for Figure 40 (fig. 3 in Sissons, 1974); and Seel House Press Ltd for Figure 9, which is based on fig. 56 in Bluck *in* Bowes and Leake, 1978.

British Geological Survey
Keyworth
Nottingham NG12 5GG
5 January 1985

G. M. Brown, FRS
Director

Contents

Illustrations

Figures		Page

Plates **Page**

Numbers in brackets refer to photographs in the BGS collections

1. Introduction

Physiography

The Midland Valley of Scotland is the name given to the relatively low lying central part of Scotland lying between the Grampian Highlands and the Southern Uplands. It is defined geologically to the north by the Highland Boundary Fault, which extends from Stonehaven in the north-east to the Firth of Clyde at Helensburgh, and its limit in the south is the Southern Upland Fault which lies parallel to the Highland Boundary Fault and extends from Dunbar to Glen App. The region has the structure of an ancient rift valley or graben in which strata between two parallel faults have subsided relative to the horst blocks on either side.

The physiographic contrast across the Highland line is a consequence of the difference in the resistance to erosion of the rocks on either side of the fault. Similarly the topography south of the Southern Upland Fault is quite distinct from that north of the fault although the line of separation makes a less noticeable feature except at the edge of the Moorfoot and Lammermuir Hills. The rocks north of the Midland Valley are the eroded remnants of the mountains formed during the later part of the Caledonian Orogeny about 400 million years ago, and the rocks of the Southern Uplands were deformed into their present configuration at about the same time. Outwith the Midland Valley Lower Palaeozoic and older rocks have been strongly folded and indurated, but between the faults the rocks are mainly of Upper Palaeozoic age and are relatively undeformed.

Although the region is known as the Midland Valley the term is appropriate only in the structural sense of a rift valley; physiographically the area is rather more diverse than the name suggests (Figure 2). Much of the region consists of farmland lying below 180 m, but there are many upland areas of rough pasture and moorland.

Thick piles of volcanic rock of Devonian and Carboniferous age, relatively resistant to erosion form a line of hills in the northern part of the region. Carboniferous lavas form the Renfrewshire Hills (522 m), the Kilpatrick Hills (401 m), the Campsie Fells (578 m) and the Gargunnock Hills (485 m), and Devonian lavas continue the north-easterly trend in the Ochil Hills (728 m) and the Sidlaw Hills (455 m). A more or less continuous strip of low ground, floored by Devonian sediments, extending from Loch Lomond to the coast around Montrose, lies between the line of volcanic hills and the Highlands. The feature widens out in the north-east to form the broad farmland of Strathmore.

Coarse conglomerates of Lower Devonian age form areas of high ground adjacent to the Highland Boundary Fault around Callander and Crieff.

In the southern part of the region the upland areas correspond to the outcrop of the older and more resistant rock-types. The conspicuous range of the Pentland Hills (579 m) consists of an upthrust outcrop of folded Silurian strata and Devonian conglomerates, sandstones and lavas. They are sharply

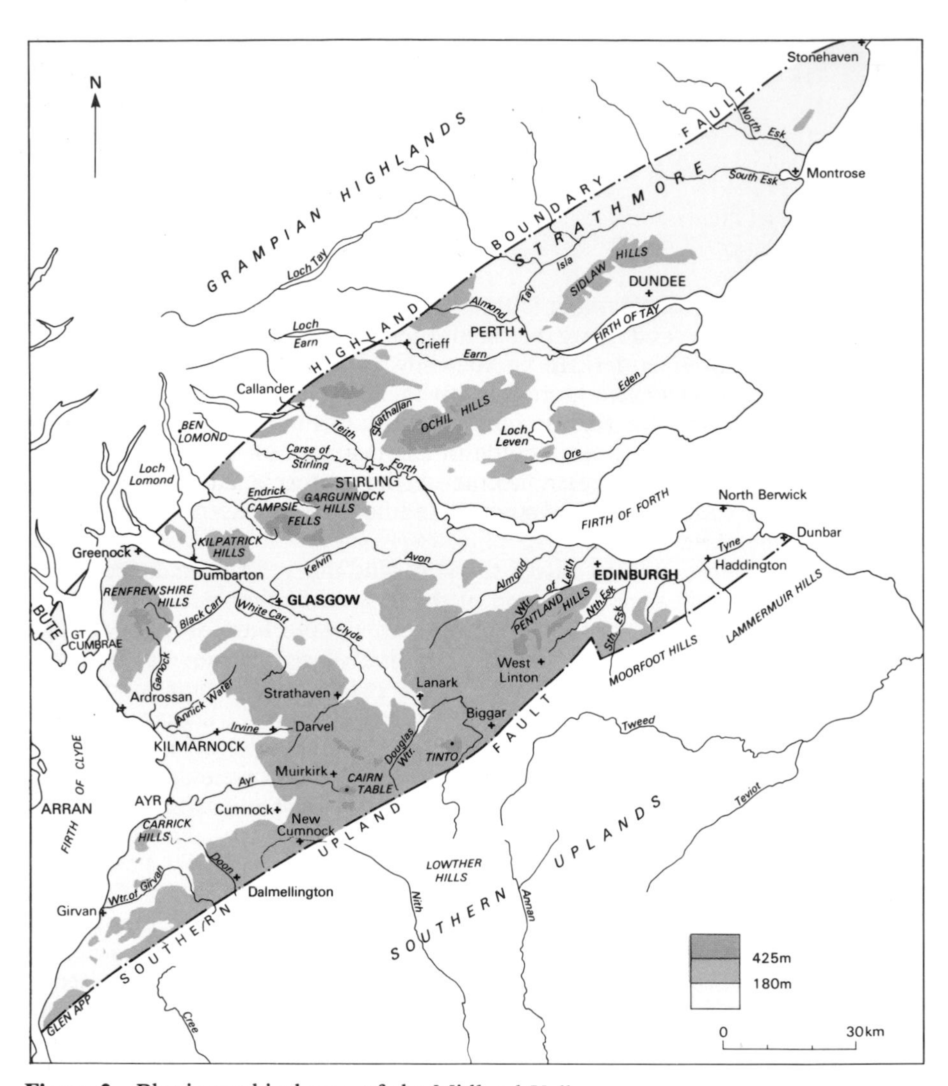

Figure 2 Physiographical map of the Midland Valley

defined on the south-east side by the Pentland Fault.

In south Lanarkshire and eastern Ayrshire a large area of dissected high ground extends from around Tinto (707 m) to New Cumnock and northwards towards Strathaven. The area contains inliers of folded Silurian rocks, which form the Hagshaw Hills (470 m) and Nutberry Hill (522 m), and intrusions of granodiorite and felsite explain the prominence of Distinkhorn (384 m), south of Darvel, and the Tinto Hills respectively. Devonian sandstones and conglomerates occupy the high ground to the south of Muirkirk which culminates at Cairn Table (593 m).

Farther south-west, the upland area north of Dalmellington is unusual in that it consists of Coal Measures strata. The relief is due to the relative resistance to erosion of dolerite sills intruded into the sediments, and sills cap Benbain (407 m), Benbeoch (464 m) and Benquhat (435 m). Devonian lavas form the Carrick Hills (287 m) south of Ayr.

The rest of the region consists of areas of undulating lowland underlain by strata of Devonian and Carboniferous age. The two largest lowland areas are central Ayrshire from Ardrossan to Ayr and extending inland to Kilmarnock and Cumnock, and the central belt from the Glasgow area to the Firth of Forth and into Fife and East Lothian. The undulating lowland landscape is enhanced by conspicuous landmarks formed of intrusive sheets or plugs of igneous rock (e.g. North Berwick Law and Traprain Law, near Haddington, in East Lothian, the castle rocks of Stirling and Edinburgh, Dumbarton Rock and Loudoun Hill, near Darvel in the west).

The present form of the landscape is the net result of several erosive agencies acting during various intervals up to a total of tens of millions of years. Attempts to chronicle the development of the landscape are necessarily tentative since glacial erosion and the mantling of the lower ground by drift deposits has removed or buried much of the evidence of the pre-Quaternary planation. The remnants of planation levels thought to be late-Tertiary in age are recorded in the clustering of summit and bench levels about a few limited altitude ranges. George (1960) recognised levels at about 700, 580 and 510 m on the northern and southern flanks of the Midland Valley and at about 325 and 180 m in central parts of the area. These he attributed to pulsatory uplift in late Tertiary times with marine erosion causing benching of the Upper Palaeozoic sediments of the rift and exposing in places the Lower Palaeozoic inliers. However, warping, differential isostatic adjustment and differing resistance to erosion of the various rock types involved make the correlation of levels problematical, and there is no sedimentary evidence to support the hypothesis of considerable submergence in mid-Tertiary times.

The drainage of the Midland Valley west of a line from Ben Lomond to a point near West Linton flows into the Firth of Clyde. East of that line the rivers run into the Firth of Forth and the North Sea. A subsidiary watershed parallel to the main watershed extends south-eastwards from near Greenock and divides the River Clyde catchment from that of the Ayrshire rivers discharging into the lower part of the Firth of Clyde.

Rivers in the northern part of the Midland Valley have their headwaters in the southern Highlands where deep glens cross the grain of the country in a south-easterly direction. They enter the Midland Valley in the strip of low ground south-east of the Highland border which extends from the Howe of the Mearns south-westwards through Strathmore and Strathallan to the Carse of Stirling. The Forth, Earn and Tay collect most of the drainage and transport it through breaches in the line of volcanic hills formed by the Campsie Fells, and the Ochil and Sidlaw Hills.

South of the Forth, the rivers draining the eastern part of the Midland Valley flow north-eastwards into the Firth of Forth.

The River Clyde rises in the central part of the Southern Uplands in the Lowther Hills and crosses into the Midland Valley near Lamington. From Lanark to the estuary the Clyde follows a north-westerly course across the south-west part of the Central Coalfield. There is a notable contrast in the geomorphological maturity of the Clyde valley above and below the confluence with the Douglas Water, near Lanark. Upstream the valley is mature, the river meanders and has a rate of fall of about 0.8 m/km but downstream it falls into a gorge at Bonnington Linn, has a rate of fall of about 15 m/km and occupies a deeply incised geomorphologically immature valley.

The upper Clyde is graded to a base-level about 170 m above OD and probably formerly drained through the Biggar Gap into the Tweed. The broad drainage basin of the lower Clyde in the Glasgow–Paisley area is abruptly constricted to a width of about 2 km between Dumbarton and Langbank where the river passes through the gap between the volcanic uplands of the Kilpatrick Hills and the Renfrewshire Hills.

Several explanations have been offered to account for the initiation and development of the drainage system. Mackinder (1902) proposed that the drainage commenced by south-easterly flowing consequent streams draining an uplifted tilted peneplain. Later Bremner (1942) and Linton (1951) suggested that the system was developed early in the Tertiary on a newly emergent cover of Cretaceous sediments mantling the older rocks and the initial easterly flowing consequent drainage was later superimposed on the underlying rocks. George (1960, 1965) argued for a mid-Tertiary submergence of an already dissected landscape and a pulsatory emergence permitting the formation of a sequence of planation platforms and benches on which the drainage system was superimposed. None of the explanations is entirely satisfactory for Scotland as a whole and especially for the Midland Valley. Difficulties in the formulation of a comprehensive account of the development of the drainage system stem from the fragmentary and tenuous nature of the evidence and the subsequent modification of it by glacial action. Additionally, the long interval of time, tens of millions of years, during which the landscape developed and the lack of any sedimentary record for much of the Tertiary and Quaternary makes explanation necessarily tentative.

The general form of the topography of the Midland Valley was established in Tertiary times, but it suffered modification during the Quaternary glaciation. Glacial erosion moulded the landscape and altered the transverse and longitudinal profiles of the valleys. Glacially eroded material was deposited mainly on the lower ground effectively obscuring the form of the solid rock surface. Relative changes in sea level led to the formation of raised beaches which are so characteristic of the coastal scenery, especially around the Firth of Clyde.

Summary of the geology

The region has the structure of an ancient rift valley with the parallel Highland Boundary and Southern Upland faults forming the limits to the area. The downfaulted strip between the two faults is floored by rocks mainly of Devonian and Carboniferous age. Comparatively small inliers of Lower Palaeozoic rocks occur on the south side of the region and rocks of Permian age occur in central Ayrshire.

Sedimentary rocks of Devonian and Carboniferous age underlie about 36 and 38 per cent respectively of the area of the Midland Valley. Igneous rocks, mainly of Devonian and Carboniferous age form about 21 per cent of the area. The geology is summarised in Figure 3.

The oldest exposed rocks are the Ordovician and Silurian sandstones, mudstones and conglomerates which occur as inliers in the Lesmahagow area, the Pentland Hills and in south Ayrshire. An upward passage from marine strata to terrestrial fluviatile rocks in the Silurian is followed by the semi-arid fluviatile clastic sediments of the Lower Devonian.

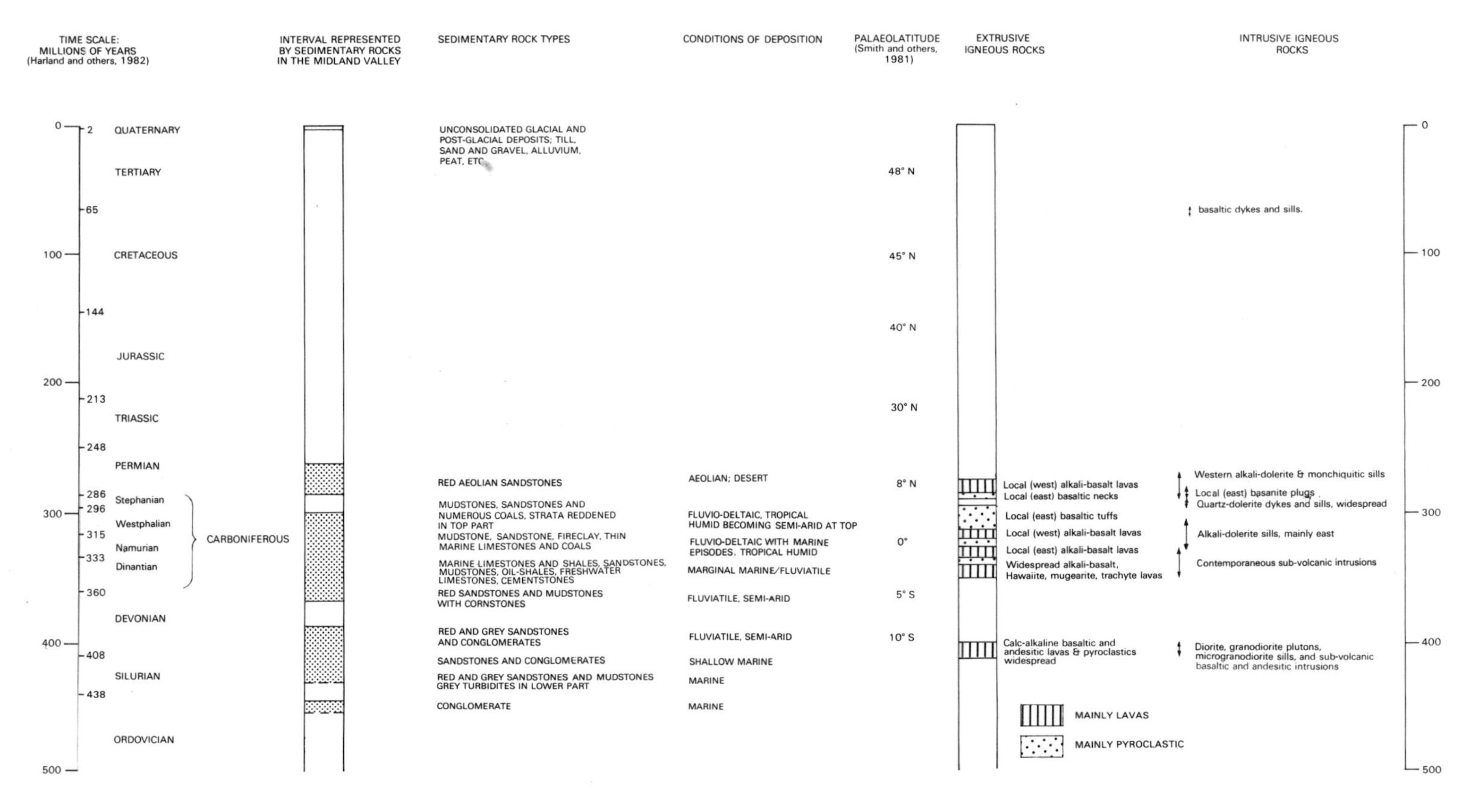

Figure 3 Summary of the geology of the Midland Valley

Great thicknesses of red and grey sandstones and conglomerates with contemporaneous piles of lava were deposited during the Lower Devonian, particularly in the north-eastern part of the area. These rocks are well exposed on the coast from the Tay estuary north to Stonehaven, and the lavas form the Ochil and Sidlaw hills.

Following deposition of the Lower Devonian rocks there was a period of uplift, folding, faulting and erosion. Important fault movements occurred on the boundary faults in Middle Devonian times and probably during the Lower Devonian. No sediments of Middle Devonian age are known in the Midland Valley.

Red sandstones and siltstones of Upper Devonian age were laid down on a peneplaned surface in the eastern and southern parts of the area, but in the west coarser sediments, the result of more vigorous erosion in an area of greater relief, accumulated in considerable thicknesses. Prolonged periods of caliche-type soil formation resulted in the development of cornstones particularly in the upper part of the division.

In the Carboniferous Period there was a climatic change to humid equatorial conditions and a large fluvio-deltaic complex of continental dimensions developed in which large quantities of sand and mud were deposited at or near sea level. Occasional flooding by the sea caused deposition of thin limestones and calcareous mudstones, and luxuriant forest growth on emergent delta-top surfaces ultimately became coal seams. The Highlands and Southern Uplands remained, in part at least, above the level of deposition. Igneous activity occurred more or less throughout the period at one locality or another and large quantities of basalt lava were extruded, particularly in the lower part of the succession.

Differential movement during the Carboniferous on fractures in the basement caused notable variation in thickness within the region, probably exerted control on the location of the igneous rocks and accounts for local unconformities and non-sequences.

Coals, ironstones, limestones and oil-shales, which have been extensively worked, formed the basis of the industrialisation of the Midland Valley during the nineteenth and first part of the twentieth centuries.

In central Ayrshire, Permian strata overlie the Carboniferous rocks and indicate that the climate had reverted to arid conditions. The strata consist of basalt lavas with intercalated sediments overlain by red wind-deposited sandstones. The dry climate of the Permian is believed to be responsible for the deep oxidation and reddening of the upper part of the Carboniferous and there is an hiatus in the sedimentary sequence between the Carboniferous and the Permian.

The Permian sandstones are the youngest sedimentary rocks present in the Midland Valley, but Triassic strata occur in Arran and floor much of the Firth of Clyde between Arran and Northern Ireland. Fragments of Lias and Cretaceous strata within the central complex in Arran and the Lias and Cretaceous outcrops in Northern Ireland suggest that Mesozoic rocks may at one time have encroached upon parts of the Midland Valley. Triassic and Cretaceous rocks also occur in the Forth Approaches.

Dykes associated with the Tertiary volcanic centres of Mull and Arran are intruded into the strata of the Midland Valley and are the most recent solid rocks in the area. Erosion during the Tertiary and glaciation during the Quaternary combined to create the present landscape.

2. Pre-Palaeozoic basement

The oldest rocks exposed within the Midland Valley are of Ordovician age but indirect evidence regarding the nature and configuration of the basement rocks may be obtained from two independent sources. A N–S deep seismic profile across northern Britain has revealed horizontal and vertical discontinuities which have been interpreted in terms of major structural elements of the crust. Samples of gneissose rocks from the basement have been carried to higher crustal levels as xenoliths in Carboniferous and Permian volcanic vents.

The seismic profile (Figure 4) suggests the presence of high-grade metamorphic basement at a depth of 7 to 9 km below the eastern Midland Valley, but at greater depths (14 to 16 km) below the adjacent Highlands and Southern Uplands structural blocks. Beneath the Highlands and Midland Valley the basement is 20 to 25 km thick and is divided by a seismic discontinuity into Upper and Lower layers. These layers cannot be recognised beneath the Southern Uplands, where the basement has different seismic characteristics and a major, lateral, crustal discontinuity may occur in the region of the Southern Upland Fault. However, similar xenolith assemblages occur in vents and intrusions on both sides of the fault indicating at least some south-eastward extension of the Midland Valley basement rocks.

Two groups of metamorphic xenoliths may be recognised, probably corresponding in broad terms to the Upper and Lower crustal divisions of the seismic profile, but also possibly intermixed in coarsely-banded gneiss formations (Plate 3.1). Quartzo-feldspathic, foliated acid gneisses, some with biotite or garnet, probably constitute most of the Upper Layer. The Lower Layer, which has seismic properties consistent with a gabbroic or dioritic composition, is probably represented by a range of pyroxene-granulites,

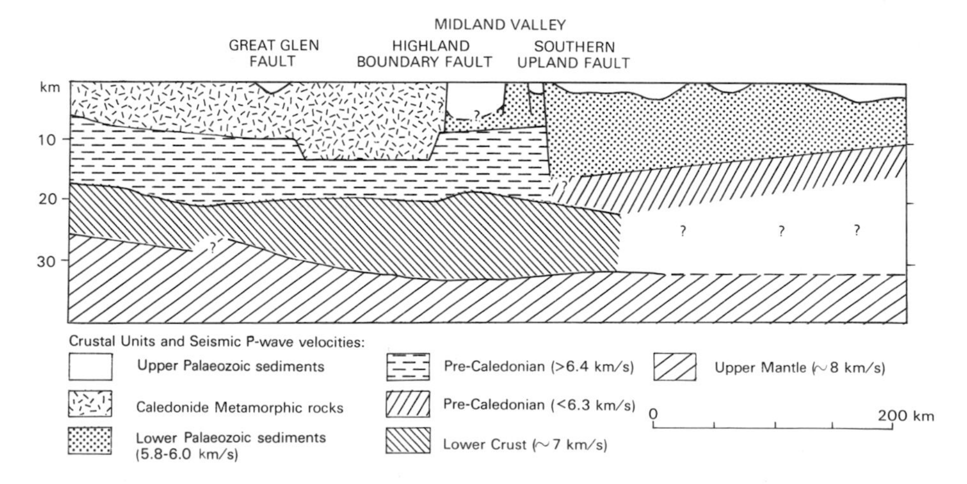

Figure 4 Schematic cross-section through the crust and uppermost mantle of northern Britain (modified after Bamford, 1979)

consisting essentially of plagioclase and clinopyroxene, usually with spinel and rarely with garnet or orthopyroxene. With increasing proportions of plagioclase a gradation occurs from basic granulites into anorthosites. Mineral assemblages, compositions and textures in both groups are indicative of granulite facies metamorphism with temperatures up to 850°C, pressures of 7 to 12 kb and inferred depths of equilibration of 20 to 35 km. Preliminary radiometric age determinations on the xenoliths suggest that the basement may have been affected by a Grenvillian metamorphic episode at 1200 to 1000 Ma. A range of unfoliated granitic rocks from tonalite to granite or trondjemite may represent later, possibly Caledonian, partial melts and intrusions within the crust.

The seismic profile shows that the base of the crust occurs at a depth of about 33 km beneath the Midland Valley. Samples of the underlying Upper Mantle material commonly occur as xenoliths in volcanic vents, often together with the less-abundant crustal material. Since they represent partial melt residues and crystal accumulations from the source regions of the Carboniferous and Permian magmas, they are discussed in Chapter 12.

3. Ordovician and Silurian

Ordovician

Rocks of Ordovician age cover a considerable area in the south-west part of the Midland Valley in the Girvan–Ballantrae district and appear also in the Craighead Inlier on the north side of the Girvan valley. An account of these appears in the volume on the South of Scotland in this series of handbooks. Reference must be made here, however, to three small lenticular inliers of Benan Conglomerate at Big Hill of the Baing, Knockinculloch (north-east of the Pilot) and Linfern Loch, which occur in the disturbed belt of sediments and volcanic rocks of Lower Devonian age that adjoins the Southern Upland Fault between Barr and Straiton. The Big Hill of the Baing outcrop, south-east of Straiton, is the largest, being five kilometres long. The conglomerate, well exposed in the Water of Girvan near Tairlaw, contains a varied assemblage of igneous pebbles derived from Arenig rocks. At Knockinculloch the conglomerate is associated with the Stinchar Limestone and Caradoc shales.

Silurian

The sediments of Llandovery and Wenlock age which occur in the Girvan–Ballantrae district and in the Craighead Inlier are described in the volume on the South of Scotland in this series of handbooks. Farther to the north-east, however, strata ranging in age from Llandovery to Ludlow or even Downtonian (Rolfe, 1973) occur in a series of inliers along the southern margin of the Midland Valley (Figure 5). These are known as the Lesmahagow, Hagshaw Hills, Carmichael, Eastfield, North Esk, Bavelaw Castle and Loganlee inliers, the last three being in the Pentland Hills. Downtonian strata also occur near Stonehaven.

As research has continued over the years, there has been a trend to recognise successively older horizons in these outcrops. The general succession in the inliers is of great importance as it records the transition from the generally marine conditions of the Llandovery, through the fish-bearing, red-bed facies of the Wenlock and Ludlow, into the true continental deposits of the Lower Devonian. Unfortunately, as the sediments become increasingly continental, so the fauna becomes more facies dependent and of less value for stratigraphical correlation with full marine successions elsewhere. Evidence of Wenlock age is present in the middle parts of the inlier successions, and a Ludlow age has been assigned to some beds at Lesmahagow (Selden and White, 1983) and Downtonian strata may also be present. Figure 6 illustrates the Silurian successions in the inliers.

Lesmahagow Inlier

The most important of the inliers of Silurian rocks in the region occurs between Lesmahagow and Muirkirk. In the inlier, the Silurian strata occur in a

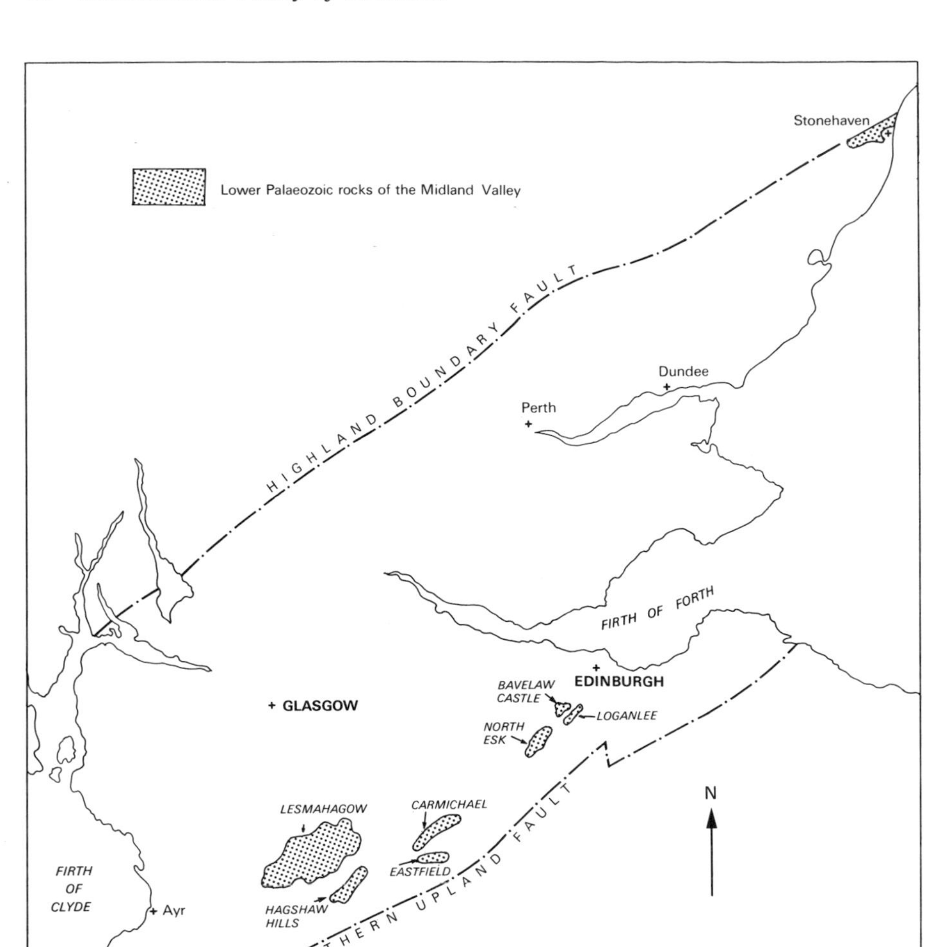

Figure 5 Location map of Lower Palaeozoic inliers of the Midland Valley

broad anticline covering a roughly oval area measuring about 23 km by 10 km. The succession, approximately 2500 m thick, is shown in Table 1.

The Priesthill Group, thought to be, at least in part, of Llandovery age, consists of grey and greenish greywackes, sandstones, siltstones and mudstones. The oldest beds are of undoubted marine origin but as the succession is traced upwards, the sediments gradually take on a more non-marine aspect. The Patrick Burn Formation at the base is a typical marine turbidite sequence of alternating thin greywackes and mudstones with subsidiary bands of laminated siltstones. The greywackes contain a redeposited shelly assemblage of brachiopods, bivalves, trilobites, including

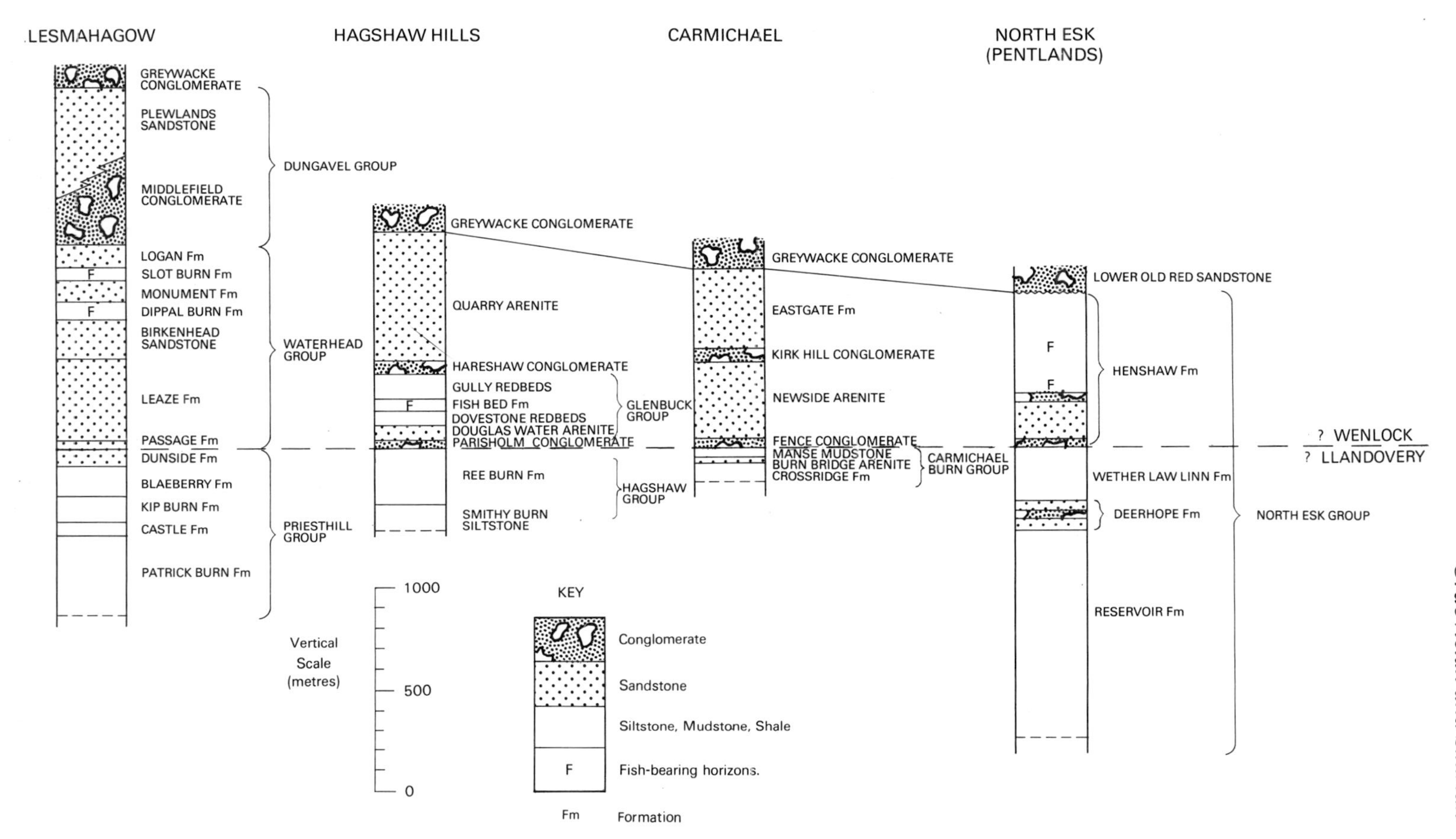

Figure 6 Silurian successions in the inliers of the Midland Valley

Table 1 Silurian succession in the Lesmahagow Inlier

Groups	Formations, etc.	Thickness (m)
Dungavel Group	Plewlands Sandstone	420
	Middlefield Conglomerate	450
Waterhead Group	Logan Formation	2–105
	Slot Burn Formation	60
	Monument Formation	50–100
	Dippal Burn Formation	55– 90
	Birkenhead Sandstone	150–190
	Leaze Formation	200–390
	Passage Formation	35– 45
Priesthill Group	Dunside Formation	40– 65
	Blaeberry Formation	75–150
	Kip Burn Formation	20–135
	Castle Formation	60
	Patrick Burn Formation	400+

Encrinurus and *Podowrinella,* ostracods and crinoid columnals. Laminated siltstones on the other hand yield an indigenous fauna comprising arthropods such as *Ceratiocaris* and *Slimonia* and the fish *Jamoytius* and *Thelodus.* The overlying unfossiliferous Castle Formation consists of massive siltstones with shale partings. Mudstones with dark laminated silty bands characterise the Kip Burn Formation, the lower part of which, the '*Ceratiocaris* beds', is highly fossiliferous yielding the arthropods *Ceratiocaris, Dictyocaris, Pterygotus, Slimonia* and the fish *Birkenia* and *Thelodus.* The upper part of the formation, the '*Pterygotus* beds', contains an abundant eurypterid fauna together with the brachiopod *Lingula* and *Ceratiocaris.* The faunas in the Kip Burn Formation reflect the start of the transition from marine to quasi- or non-marine conditions in the inlier. Mudstones again predominate in the succeeding Blaeberry Formation, including some highly fossiliferous bands containing *Lingula*, the gastropod '*Platyschisma*', *Slimonia* and ostracods. The Dunside Formation, at the top of the Priesthill Group, consists of flaggy micaceous sandstones with some cross-bedding. These strata, the '*Trochus* beds', contain abundant specimens of '*Platyschisma helicites*'.

The Waterhead Group, of Wenlock/Ludlow age, comprises red, green and variegated beds including sandstones, siltstones, mudstones and shales and is unfossiliferous except for two fish-bearing horizons. The group as a whole is thought to represent deposition in shallow marine, deltaic and lagoonal conditions. The Passage, Leaze, Monument and Logan formations are composed of fine- to coarse-grained sandstones and mudstones which may be greenish grey or red-brown. Mudcracks, indicative of subaerial conditions, are common in the red beds but are not found in the green beds whereas occasional *Lingula* are found only in the latter in the lower two formations. The Birkenhead Sandstone, of deltaic origin, is distinguished by its massive, orange, cross-bedded sandstones containing pebbles of acid igneous rocks. The fish-bearing horizons occur in the Dippal Burn and Slot Burn formations which consist of greenish grey mudstones, siltstones and thick sandstones. The

Plate 1 Silurian fossils

1 *Pterygotus bilobus.* **2** *Eoplectodonta penkillensis*, ×2. **3** *Acernaspis (Eskaspis) sufferta*, ×2. **4** *Coolinia applanata.* **5** *Thelodus scoticus.*

fossils are found in dark grey laminated siltstones which alternate with grey mudstones. Fossils obtained include unidentifiable plants, arthropods and fish including *Birkenia, Lasanius* and *Thelodus*.

The Dungavel Group is unfossiliferous but is thought to be of Ludlow or even Downtonian age. The oldest beds are the conglomeratic basal members of the Middlefield Conglomerate which contain pebbles and boulders up to 45 cm in diameter consisting dominantly of quartzite, vein-quartz, jasper and some acid igneous rocks. Lenses of coarse-grained, greyish brown, feldspathic sandstones alternate with the conglomeratic units. As the succession is ascended, the sandstone becomes the dominant member, enclosing isolated conglomeratic lenses. The greyish brown, micaceous, cross-bedded fluviatile sandstones of the Plewlands Sandstone are the youngest rocks in the inlier. The succeeding 'Greywacke Conglomerate' has been taken as the local base of the Devonian in the area.

Hagshaw Hills Inlier

The steeply inclined strata of this inlier form an overturned asymmetrical anticline which can be traced from the vicinity of Little Cairn Table north-eastwards to Rob's Hill, just west of Douglas (Figure 5). The inlier is about 34 km^2 in area and the succession present, about 1500 m thick, has been subdivided as shown on Table 2.

Table 2 Silurian succession in the Hagshaw Hills Inlier

Groups	Formations, etc.	Thickness (m)
	Quarry Arenite	600
	Hareshaw Conglomerate	75
Glenbuck Group	Gully Redbeds	120
	Fish Bed Formation	60
	Dovestone Redbeds	75
	Douglas Water Arenite	75
	Parishholm Conglomerate	40
Hagshaw Group	Ree Burn Formation	280
	Smithy Burn Siltstone	120

The Hagshaw Group consists of dark to medium-grey greywackes, siltstones, mudstones and shales of marine origin with, in the Ree Burn Formation, a poorly preserved shelly fauna including the brachiopods *Protochonetes* aff. *edmundsi, Howellella*, various bivalves, ostracods and species of *Encrinurus* and *Podowrinella*. An earlier record of the graptolite *Monoclimacis vomerina* has not been confirmed but the above fauna is thought to indicate an Upper Llandovery or Lower Wenlock age for the Hagshaw Group.

The Parisholm Conglomerate is a pebble and cobble conglomerate with a varied suite of igneous and sedimentary clasts including felsite, porphyrite, spilite, keratophyre, quartzite, vein-quartz, greywacke, mudstone, chert and ?tuff. From its composition it may be tentatively correlated with the Fence Conglomerate of the Carmichael Inlier which has a similar pebble suite.

The succeeding formations of the Glenbuck Group consist of grey and greyish red sandstones, calcareous mudstones, siltstones and shales with a fish fauna, including species of *Birkenia, Lasanius* and *Lanarkia*, from the Fish Bed Formation.

They are succeeded by the Hareshaw Conglomerate, a pebble and cobble conglomerate containing an abundance of vein-quartz and quartzite clasts as well as pebbles of porphyrite, granite, rhyolite, ?gneiss, schist, mudstone, chert and pyroclastics.

The highest formation seen in the inlier is the Quarry Arenite, a greenish grey and pale red, medium- and coarse-grained sandstone often containing intraclast fragments of red shale. Sedimentary features such as current bedding and scour-and-fill structures are common throughout the formation.

Carmichael Inlier

This inlier occupies a strip of ground about 11 km long by 2 km wide on the north-west side of Tinto (Figure 5). The sequence comprises a lower succession of mainly greenish grey mudstones and siltstones (Carmichael Burn Group) succeeded by a coarser-grained sequence divided into the Fence Conglomerate, with igneous-rock pebbles, the Newside Arenite, the Kirk Hill Conglomerate, with quartzite pebbles and the Eastgate Formation. Many of the lithologies can be matched quite closely with rocks from parts of the Pentland Hills and Hagshaw Hills inliers as well as from the Blair–Knockgardner–Straiton inlier at Girvan.

In the lower part of the Carmichael Burn Group a relatively large fauna was obtained from thin siltstone seams within a mudstone sequence. The fauna includes annelids, brachiopods, cephalopods, ostracods, phyllocarids (*Ceratiocaris*), crinoid columnals and graptolites (mainly monograptids of the *vomerina* type) suggesting a late Llandovery age for the lowermost beds of the inlier (Rolfe, 1960).

Eastfield

A small inlier, about 5 km long and 1 km wide, is present on the south side of Tinto near Eastfield Farm. The sediments are poorly exposed and consist of greenish sandstones with conglomerate bands containing pebbles of quartzite, chert, jasper, vein-quartz, granite, porphyry, etc. No fossils have so far been found in these beds and the age of the rocks is based on lithological criteria.

Pentland Hills

Silurian strata crop out in three small areas in the Pentland Hills, where they form the North Esk, Bavelaw Castle and Loganlee inliers. They are everywhere highly inclined and are overlain with marked unconformity by beds of Lower or Upper Devonian age.

North Esk Inlier

This inlier, occupying about 6 km^2 of country at the head of the Lyne Water and North Esk River, is the largest area of Silurian rocks in the Pentland Hills.

The succession begins in the river below the North Esk Reservoir with the alternating purple and grey mudstones and siltstones of the Reservoir Formation. The beds are sparsely fossiliferous, with brachiopods such as *Craniops* and *Glassia* found scattered throughout the sequence. In the upper part of the formation a benthonic fauna is found including the brachiopods *Atrypa, Coolinia, Leptaena, Resserella*, the trilobites *Acernaspis* and *Harpidella*, ostracods and crinoid columnals. In the same part of the succession certain flaggy sandstones in the Gutterford Burn contain the well known eurypterid and starfish beds. Rare graptolites obtained from the Reservoir Formation include *Koremagraptus* and *Monoclimacis* cf. *vomerina*.

The Reservoir Formation is succeeded by the more arenaceous sediments of the Deerhope Formation consisting of flaggy greywacke interbedded with purplish grey and green siltstones and mudstones. These grade upwards into the coarse greenish grits and pebbly sandstones referred to as the Haggis Grit and Conglomerate. Silty beds in various parts of the formation have yielded a coral and shelly fauna including the coral *Pleurodictyum*, the brachiopods *Chonetes* and '*Strophomena*' and the bivalves *Modiolopsis* and *Orthonota?*. Trilobites and ostracods also occur.

The Haggis Grit and Conglomerate is overlain by highly fossiliferous greenish brown mudstones and siltstones of the Wether Law Linn Formation. These beds have provided many species of brachiopods, bivalves and gastropods as well as a few cephalopods and trilobites. Prominent species include brachiopods such as *Skenidioides lewisii* and *Eoplectodonta penkillensis*, species of *Acernaspis, Encrinurus, Podowrinella*, and the ostracod *Craspedobolbina (Mitrobeyrichia) impendens*. A 250-mm bed of white clay, thought to be a volcanic ash, has been recorded between 3 and 10 m above the base of the formation.

The Reservoir, Deerhope and Wether Law Linn formations were formerly considered to be of Wenlock–Ludlow age, but further research suggests a late Llandovery age.

Both in the Lyne Water and in the North Esk, the Wether Law Linn Formation is succeeded by a red igneous-pebble conglomerate followed by red mudstones, siltstones, sandstones and intercalated chert-pebble conglomerates. These red beds form the Henshaw Formation and contain a restricted fauna including bryozoa, eurypterid fragments, *Ateleaspis, Birkenia* and *Lasanius*. The fauna was formerly placed in the Downtonian but recent work suggests that the Henshaw Formation is, at least in part, of Wenlock age.

Bavelaw Castle

This inlier occupies a triangular area extending eastwards from Bavelaw Castle to the western slope of Black Hill and south to the northern slope of Hare Hill. The beds consist of grey to purplish grey mudstones or silty mudstones with interlaminated siltstones and occasional thin flaggy sandstones. In lithology they resemble the lowermost beds of the North Esk Inlier, but the fauna is slightly more varied and includes *Glassia*, species of *Dictyocaris* and *Orthoceras* as well as a monograptid and *Retiolites geinitzianus*. These beds were considered by Peach and Horne (1899) to be of Wenlock age but a detailed re-examination of the fauna led Lamont (1947) to claim that they are older and of Llandovery age.

Loganlee

This inlier forms a narrow belt of country extending from near Loganlee along the eastern side of Black Hill to the west slope of Bell's Hill. The sediments are best exposed on the south-east slope of Black Hill close to the footpath leading from Loganlee to Bavelaw. They consist of greenish grey and purple shales with layers of siltstone and occasional thin beds of flaggy sandstone as well as a number of thicker sandstone beds. The shales have yielded several species of graptolites which were once considered to represent a Wenlock assemblage. More recent work, however, has failed to confirm these findings and, by analogy with the latest work on the nearby North Esk Inlier, the beds could well be of Llandovery age.

Downtonian of Stonehaven

The Downtonian of this area consists of shales, mudstones and sandstones of the Stonehaven Group which rest unconformably on the Cambro–Ordovician rocks of the Highland Border Series. They are succeeded, apparently without significant unconformity, by massive conglomerates of the Dunnottar Group. The steeply dipping beds, exposed in nearly continuous section on the shore near Stonehaven, are divisible into two formations.

The lower division, the Cowie Formation, has a total thickness of about 730 m and consists of a thin basal breccia followed by dull red, grey and yellow sandstones with numerous intercalations of red or grey mudstone. The sandstones are fine- to medium-grained, cross-bedded and contain abundant metamorphic rock fragments. In the upper part of the Cowie Formation, conglomerates containing rounded pebbles of acid volcanic rocks are succeeded by grey sandstones and shales containing *Dictyocaris slimoni* at intervals throughout their thickness as well as in the important Cowie Harbour Fish-bed near their base. The latter has also yielded the arthropods *Archidesmus, Ceratiocaris, Hughmilleria norvegica, Kampecaris?* and *Pterygotus* and the fish *Hemiteleaspis heintzi, Pterolepis* and *Traquairaspis campbelli*.

The Carron Formation forms the upper division of the Stonehaven Group and consists of about 820 m of medium- to coarse-grained, cross-bedded reddish brown sandstones with a high proportion of acid volcanic debris. Locally the beds are conglomeratic, containing pebbles of a variety of igneous rocks and metasediments while continuing sporadic volcanic activity is demonstrated by an agglomerate, containing angular fragments of biotite- and hypersthene-andesite, which is intercalated with sandstones on the south side of Stonehaven harbour.

4. Devonian

The Devonian rocks were laid down in an internal basin within a large continent. The basin lay between the mountains of the Caledonian Orogeny in the north and the newly formed Southern Uplands in the south. The sediments are of continental facies and are associated with extensive piles of andesitic and basaltic lavas. Vigorous erosion in the source areas to the north and south of the basin provided abundant detritus which was deposited in mountain-front fans and farther distributed by braided-river systems.

The system is subdivided into Lower, Middle and Upper Devonian, but in the Midland Valley the Middle Devonian is not known to occur. The term Old Red Sandstone has been applied to Devonian rocks of continental facies and distinguishes them from the marine facies of south-west England and Europe.

The Lower Devonian rests with angular unconformity on the underlying rocks except in the Lesmahagow Inlier and at Stonehaven where the Devonian rocks follow the Silurian rocks without any apparent discordance.

A period of folding, faulting and erosion preceded deposition of Upper Devonian sediments.

The base of the Devonian succession is marked by an unconformity or, in successions where there is no apparent discordance at the base, it is taken at a lithologically convenient horizon. The fossils present are inadequate to determine palaeontologically the Silurian–Devonian boundary. Radiometric age-dating of Lower Old Red Sandstone lavas suggests that some of them may be of Silurian age. The Upper Devonian rests on a major unconformity and passes up without apparent break into sediments of Carboniferous age. The Upper Devonian boundary cannot yet be precisely identified palaeontologically in Scotland, but it is placed at a stratigraphically convenient horizon between beds containing fossil fish of Upper Devonian age and strata proven to be Carboniferous. The difficulty arises from a major but transitional and probably diachronous facies change from continental to deltaic/marine conditions which occurred probably in the early part of the Carboniferous.

The Devonian rocks of the Midland Valley contain a fauna consisting of fish, rare arthropods and a flora which has yielded few good fossil plants but some assemblages of spores have been recorded which are valuable in broad scale correlation. The fauna is peculiar to a terrestrial environment and direct comparison with marine Devonian successions elsewhere is very difficult. The general correlation of the Midland Valley succession with marine Devonian successions is shown on Figure 7.

The palaeolatitude of the Midland Valley in the Lower Devonian and its position relative to the limits of the marine and continental facies of the Devonian are shown in Figure 8.

Lower Devonian

The Lower Devonian sediments were laid down in most areas on a surface of

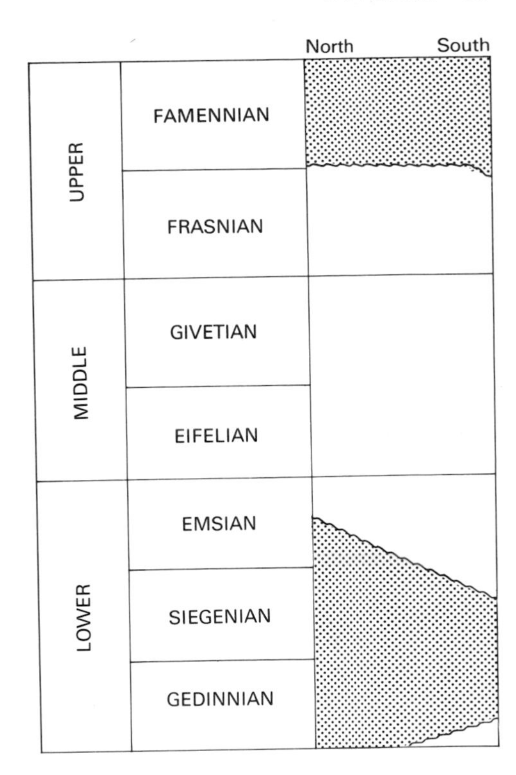

Figure 7 Devonian stages and their representation in the Midland Valley

folded and eroded Lower Palaeozoic sediments and are molasse deposits which resulted from rapid erosion in the upland regions north and south of the main depositional area. The eroded material was washed down into the valleys, transported and deposited in the first instance as coarse valley-fill deposits within the mountains. The lighter fractions were carried farther and laid down in coalescing alluvial fans at the mountain front and more distally on the flood plains of braided and meandering rivers and in shallow lakes.

Andesitic, basaltic and more rarely rhyolitic lavas are intercalated with the sedimentary succession in many places. These are described in the next chapter. Contemporaneous erosion of the lavas provided a considerable proportion of the detritus incorporated into the sediments, especially in the southern part of the area.

One possible model of Lower Devonian sedimentation consists of two south-westerly flowing river systems, one in the northern part of the region and the other in the south, separated by volcanic uplands in the central part of the Midland Valley. Drainage from the mountains resulted in the formation of coarse piedmont fans in belts at the margins of the Midland Valley with a system of braided streams and flood plains draining the area towards the south-west (Figure 9). In Kincardineshire there is evidence that the sediment was derived from the north-east and possibly the basin closed to the north-east. It has also been suggested that the main centre of deposition migrated towards the south-west with time, controlled by hinge movements on the Highland Boundary Fault (Bluck, 1978).

The outcrop distribution of the Lower Devonian rocks is shown in Figure 10. The major controls on the outcrop, but not on the original extent of deposition, are the Highland Boundary Fault and the Ochil Fault in the north,

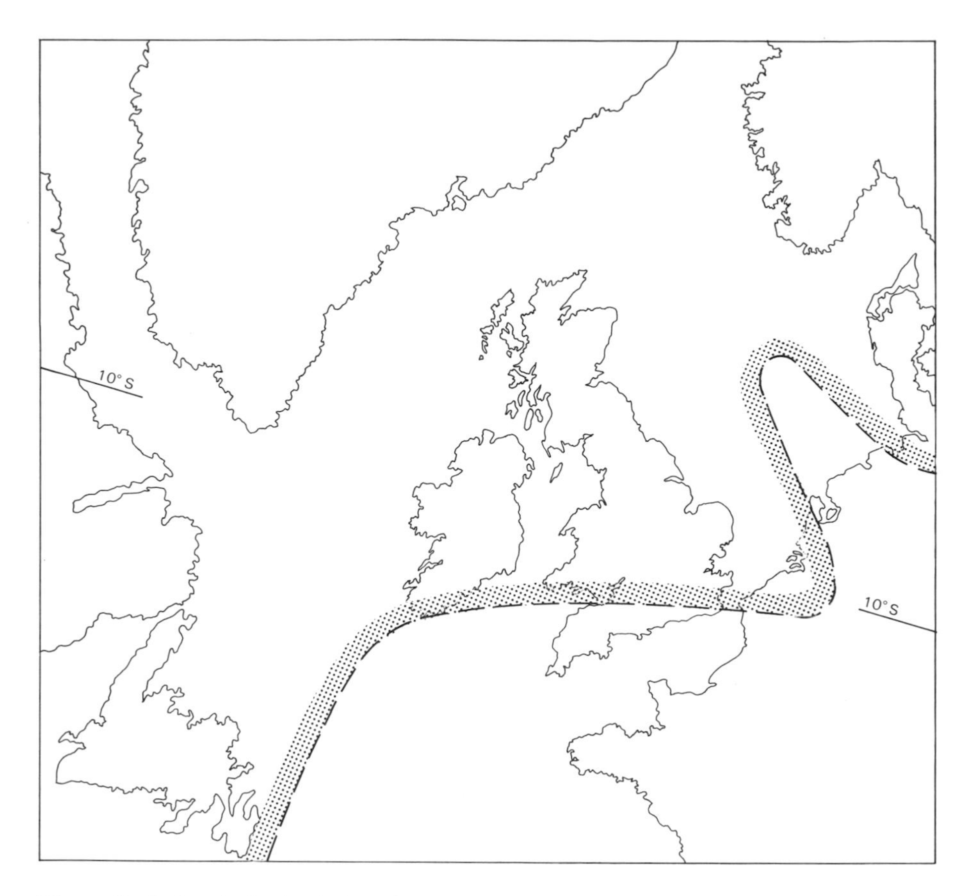

Figure 8 Palaeolatitude during the Lower Devonian (after Smith and others, 1981) and approximate limit of continental facies

and the Southern Upland Fault and its associated faults in the south.

Fossils are not common in the Lower Devonian rocks but interesting material has been found in several fish-beds, notably in the Forfar–Letham–Brechin area. The fauna includes the arthropods *Kampecaris* and *Pterygotus* and the fish *Cephalaspis, Climatius, Mesacanthus* and *Pteraspis*. The plants include *Arthrostigma, Parka, Psilophyton* and *Zosterophyllum*. Spore assemblages from various horizons have given some indication of the ages of the beds sampled.

Northern part of the Midland Valley

The Devonian strata in the northern part of the Midland Valley are folded into a broad syncline and anticline with their axes roughly parallel to the Highland Boundary Fault. The Strathmore Syncline, which lies to the north-west of the Sidlaw Anticline is asymmetrical. The north-west limb of the syncline is steep and locally overturned in the vicinity of the Highland Boundary Fault, but the south-east limb is gently inclined. The axis of the syncline extends from Stonehaven south-westwards to Loch Lomond and the Clyde Estuary.

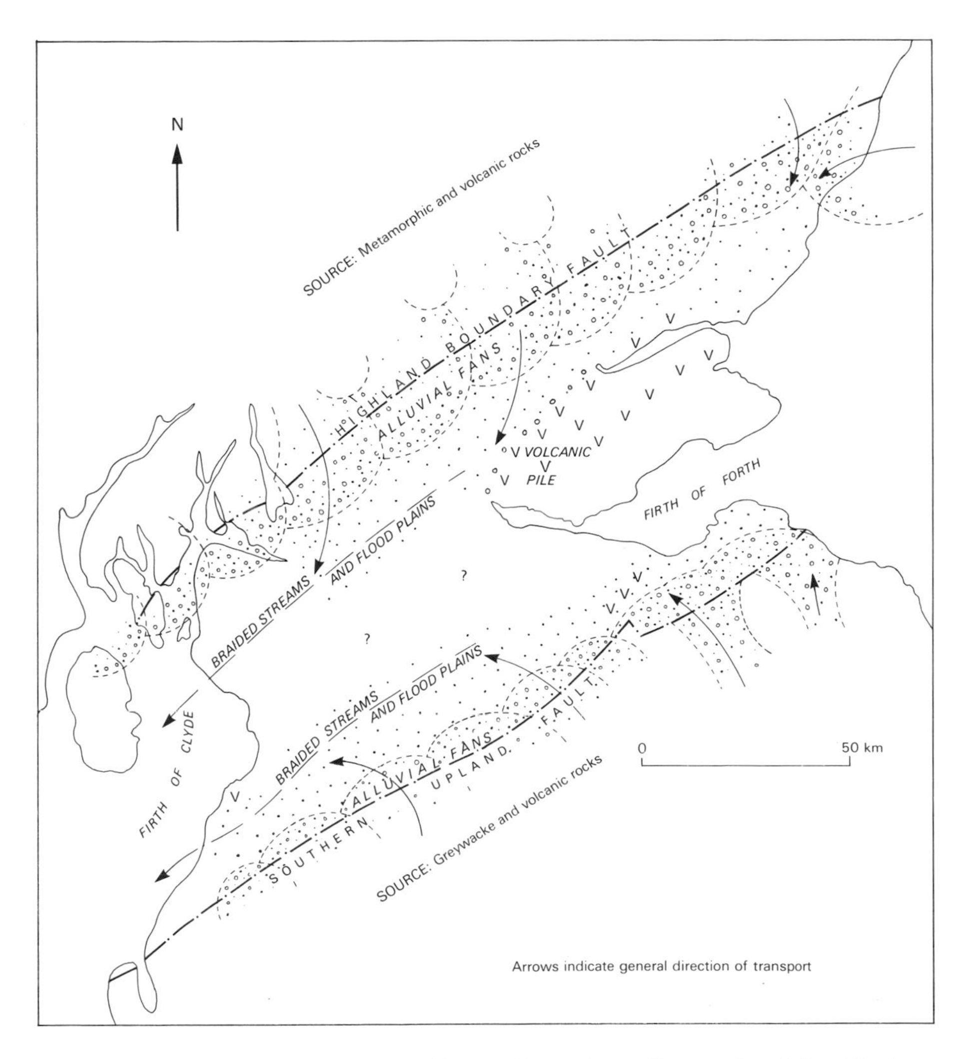

Figure 9 Palaeogeographical map of Lower Devonian sedimentation (after Bluck, 1978)

The Sidlaw Anticline is almost symmetrical and its axis runs from Montrose south-westwards to meet the Ochil Fault near Tillicoultry.

The subdivisions of the Lower Devonian in the Strathmore area are given in Figure 11, the total sequence being about 7500 m thick. This figure is reduced to about 4000 m in the Callander–Dunblane area where the lowest two subdivisions have been cut out by overlap. The lateral extent of the basal two subdivisions along the axis of the basin is unknown.

Dunnottar Group

The Dunnottar Group is the oldest of the Lower Devonian subdivisions and is well exposed on the coast south of Downie Point, near Stonehaven.

The Group is subdivided as follows:

		m
Tremuda Bay Volcanic Formation	Olivine-basalt lavas	60
Dunnottar Castle Conglomerate	Coarse conglomerate mainly of metamorphic rocks	1035
Strathlethan Formation	Grey sandstone with andesitic lava and agglomerate	350
Downie Point Conglomerate	Coarse conglomerate mainly of metamorphic rocks	170

The basal formation, the Downie Point Conglomerate, rests with an erosive base, but without apparent angular discordance on the Stonehaven Group of the Silurian. The conglomerate consists of well-rounded boulders, up to 0.6 m across, mainly of metamorphic rocks including quartzite and schistose grit. This contrasts with the content of the conglomerates in the underlying Stonehaven Group which are predominantly acid volcanic rocks.

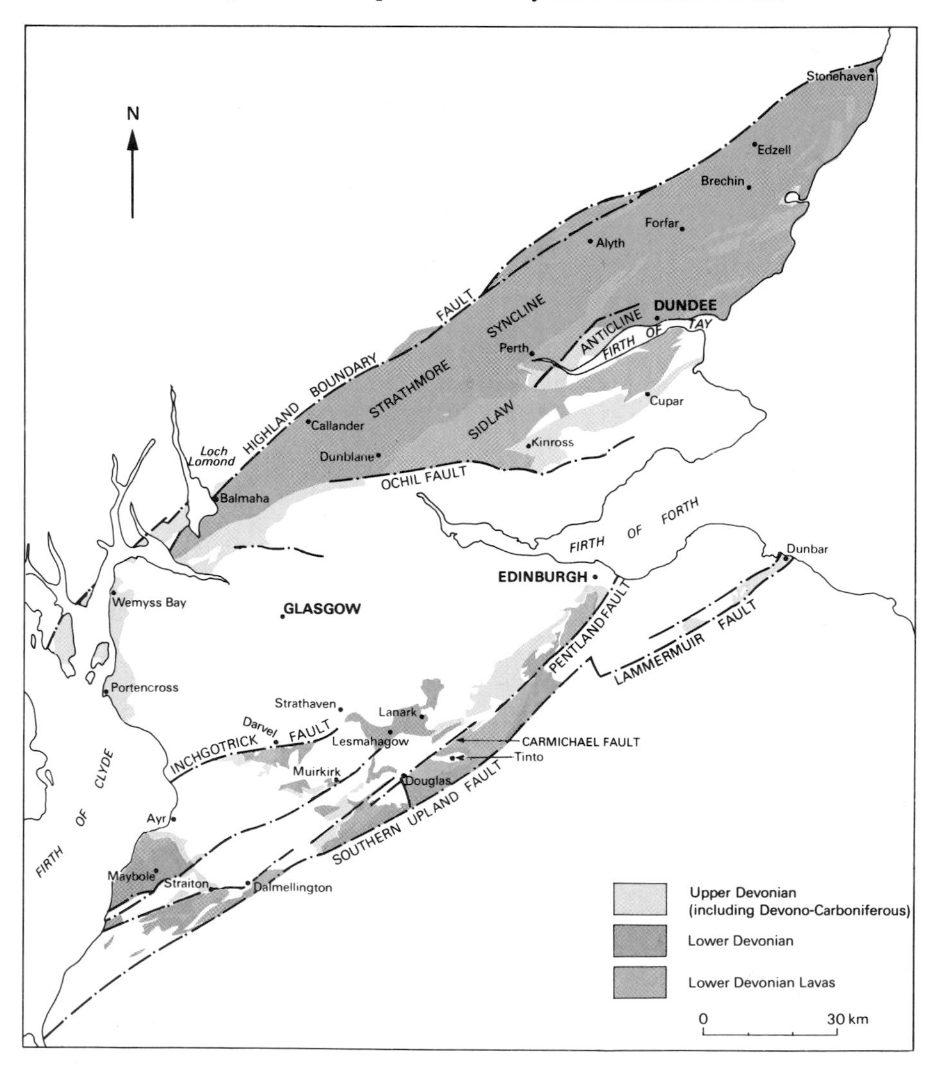

Figure 10 Outcrop distribution of Devonian rocks in the Midland Valley

The remainder of the Dunnottar Group is mainly very coarse conglomerate with boulders up to 1 m across of quartzite and schist. The top part of the group consists of olivine-basalt lavas, but the full thickness of these is not known.

Crawton Group

The Crawton Group rocks are best exposed in coastal sections north and south of Inverbervie.

The rocks consist principally of conglomerates of metamorphic rocks and locally of volcanic detritus. The maximum development is seen on the coast south of Stonehaven where the following subdivisions can be seen:

		m
Crawton Volcanic Formation	Distinctive olivine-basalt or basaltic andesite lavas	0 – 30
Whitehouse Conglomerate	Mainly coarse conglomerate of metamorphic rocks	60 – 300
Gourdon Formation	Mainly coarse dull and pebbly sandstones	150 – 365
Rouen Formation	Mainly coarse conglomerate of metamorphic rocks	185 – 215

The Crawton Volcanic Formation is known only in the north-eastern part of the area, but an approximate correlation has been made with a 'quartz porphyry' at Glenbervie and the 'Lintrathen Porphyry' near Alyth which have been shown to consist of ignimbrite. The Lintrathen Porphyry has given a radiometric age (411 ±6 Ma) which places it close to the Silurian/Devonian boundary on the current radiometric time scale. Near Dunkeld the ignimbrite rests directly on Dalradian rocks indicating that the Dunnottar Group and most of the Crawton Group as seen to the north-east have been overlapped in this part of the Highland border area by the uppermost part of the Crawton Group.

Arbuthnott Group

The Arbuthnott Group occurs on the north side of the Strathmore Syncline between Stonehaven and Loch Lomond, on the south side of the syncline from Stonehaven south-eastwards to the Ochil Fault at Tillicoultry and on the south-east limb of the Sidlaw Anticline.

The group is subdivided into four strikingly diachronous formations including coarse conglomerates at the Highland Border which are considered to be the lateral equivalents of the volcanic formations of the Ochil and Sidlaw Hills. The relationships are shown on Figure 11.

The Johnshaven Formation, in the north-east of the area, consists of conglomerates made up of metamorphic and igneous rocks. At its maximum it is about 1800 m thick. Its lateral equivalent further to the south-west is the Dundee Formation which is up to 1500 m thick. It consists of medium- and coarse-grained, cross-bedded sandstones with intercalations of flaggy lacustrine sandstones which were formerly an important source of paving stones and roofing tiles.

The Montrose Volcanic Formation consists of several lenticular masses of olivine-basalt and andesitic lava interbedded with the Johnshaven and Dundee formations.

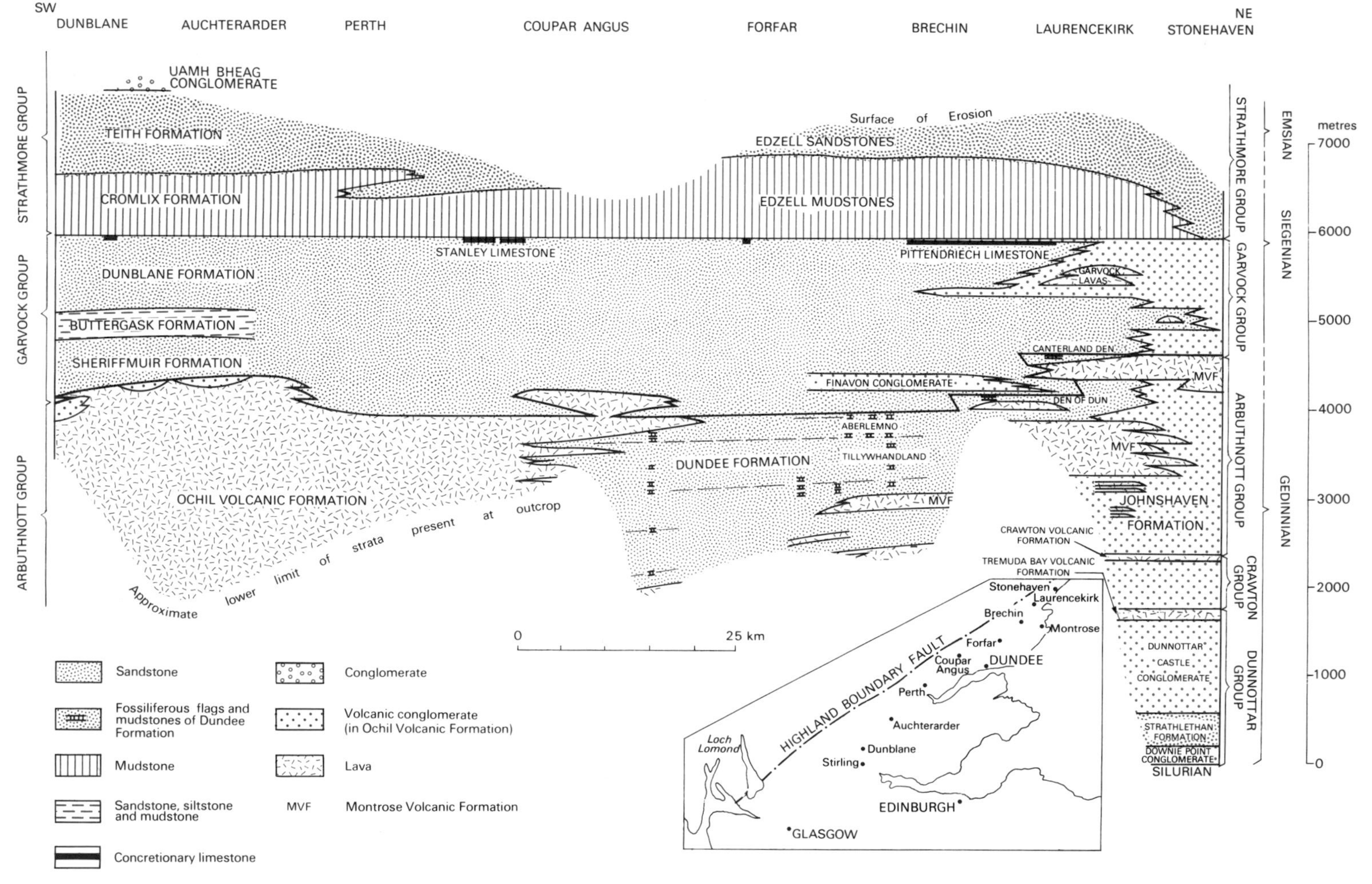

Figure 11 Lateral variation in the Lower Devonian on the SE limb of the Strathmore Syncline

In the Ochil Hills the Arbuthnott Group consists almost entirely of andesitic and basaltic lavas with beds of volcanic detritus comprising the Ochil Volcanic Formation which may be up to 3000 m thick.

There are a number of important fish-bearing horizons within the lacustrine flagstones particularly in the Forfar, Letham and Brechin area. These include the Aberlemno, Tillywhandland, Canterland Den and Den of Dun fish-beds.

Garvock Group

The outcrop of the Garvock Group occurs on both limbs of the Strathmore Syncline from near Edzell to Loch Lomond on the north side, and from Stonehaven to Bridge of Allan on the southern limb. It also forms an outcrop on the south limb of the Sidlaw Anticline on the coast north and south of the Tay estuary.

Around Laurencekirk the strata consist of conglomerates of both metamorphic and volcanic rocks, and lava flows of basalt and andesite occur at several horizons. In the Forfar–Brechin area, the conglomerate is largely replaced by brown, cross-bedded sandstones. Calcareous detritus of intra-formational origin occurs in the sandstones and there is a thin but persistent concretionary limestone near the top of the group in the Brechin area which is possibly equivalent to similar limestone occurrences at Stanley, near Perth, and at Dunblane. The limestone is a caliche type of deposit which is sufficiently persistent laterally to be of value as a stratigraphical marker.

Strathmore Group

The Strathmore Group occupies the core of the Strathmore Syncline and its outcrop extends from Laurencekirk to the Firth of Clyde.

In the north-east, around Edzell, the group consists of a lower formation called the Edzell Mudstones and an upper formation, the Edzell Sandstones. The mudstone formation is estimated to be about 1200 m thick and consists of red and green mottled, poorly bedded mudstone with beds of sandstone and siltstone. The sandstone formation consists of pale red sandstone, pebbly in places and up to 600 m thick.

On the north-west side of the syncline a thick group of coarse conglomeratic sediments called the Gannochy Formation intervenes between the Edzell Sandstone and the Edzell Mudstones and largely replaces them.

In the Dunblane and Callander area the stratigraphy is broadly similar. A lower mudstone and siltstone formation called the Cromlix Formation represents the Edzell Mudstones and is overlain by the Teith Formation which is predominantly arenaceous.

On the north-west of the syncline, around Callander, the sediments are predominantly arenaceous, but at Uamh Bheag, north-east of Callander, the sandstones are overlain by conglomerates containing clasts of both metamorphic and volcanic rocks.

Western Midland Valley

In the Loch Lomond area, the Lower Devonian rocks have been correlated with the Strathmore, Garvock and Arbuthnott groups of Strathmore and are believed to be about 800 m thick (Morton, 1979). However an interpretation of the gravity measurements suggests a thickness in the range 1500 to 1800 m (Qureshi, 1970).

The oldest Lower Devonian rocks seen in the area belong to the Balmaha Conglomerate which consists of cobbles and pebbles of quartzite in the lower part and pebbles of quartzite and lava higher up. The conglomerates are overlain by sandstone with beds of conglomerate.

At Portencross, on the north Ayrshire coast, the oldest sediments are the Sandy's Creek Beds which are tectonically disturbed grey sandstones and siltstones. Spores from these beds indicate a Lower Devonian age but the beds could be older.

The younger Portencross Beds, which are about 450 m thick, consist of brownish sandstones with lenticular, irregular intercalations of conglomerate containing pebbles of quartzite mainly, with lesser quantities of chert, sandstone and lava.

Southern part of the Midland Valley

In the much faulted tract in the south of the region, Lower Devonian rocks occur in a number of outcrops which extend from the Pentland Hills in the east to Dalmellington and Maybole in the south-west. The detailed stratigraphy of the rocks is not well known.

In the Pentland Hills the Lower Devonian strata rest with angular unconformity on Silurian rocks. The sediments consist of about 600 m of coarse greenish grey conglomerates and coarse pebbly sandstones. The content of the conglomerate consists of pebbles of greywacke, chert and jasperised lava and they tend to be coarser towards the base of the succession. The sediments thin towards the north-east where they interdigitate with and are replaced by a sequence of lavas and tuffs up to 1800 m thick, varying in composition from basalt to rhyolite. The lavas are probably equivalent in age to part of the sedimentary sequence, but there is no palaeontological evidence for the age of these rocks.

The outcrop continues to the south-west in a strip between the Southern Upland Fault and the Carmichael Fault. The Lower Devonian rocks in this area can be subdivided into three groups. The Greywacke Conglomerate at the base is overlain by a group of lavas which in turn is succeeded by another group of sandstones and conglomerates.

The Greywacke Conglomerate is exposed north and south of Tinto Hill where it is very coarse. In addition to greywacke it contains pebbles of quartz, jasper and chert.

The lavas consist mainly of basalts and andesites with some trachytes, but a rhyolite breccia occurs near West Linton. Lavas are present in the succession along the entire length of the outcrop from West Linton to Corsencon Hill, near New Cumnock.

The strata overlying the lavas consist of purplish and green micaceous feldspathic sandstones and conglomerates containing clasts of volcanic rock. The youngest preserved component of the sequence is a very coarse volcanic conglomerate called the Dungavel Conglomerate. The sandstones and conglomerates occur in two outcrops contained in a syncline parallel to the Southern Upland Fault.

In the Lanark area to the north-west of the Carmichael Fault, the Greywacke Conglomerate rests without apparent discordance on the Silurian rocks. It is about 450 m thick in the southern part of the area and it thins rapidly north-westwards to about 8 m west of Lesmahagow. The conglomerate

Plate 2 Devonian fossils

1 *Bothriolepis hicklingi.* **2** *Holoptychius nobilissimus* (scale). **3** *Mesacanthus mitchelli.*
4 *Arthrostigma gracile.* **5** *Cephalaspis powriei.*

passes up into medium- and coarse-grained sandstones with sporadic pebbly horizons. There are no lavas in the Lower Devonian outcrop in the Lanark area, north-west of the Carmichael Fault. The Greywacke Conglomerate and the overlying sandstones are probably equivalent to the sediments below the lavas elsewhere and the lavas have either been completely eroded or were never present in this area.

Lavas reappear in the sequence south of Darvel. The sediments below the lavas form an outcrop around Distinkhorn where they consist of red and purplish, cross-bedded feldspathic sandstone with a conglomerate at the base and are about 670 m thick. The content of the conglomerate is mainly greywacke, but there are also pebbles of quartz, quartzite, chert, jasper and acid igneous rocks. The conglomerate rests on Silurian rocks with apparent conformity.

In south-west Ayrshire there are two main areas of outcrop of Lower Devonian rocks. The Maybole outcrop extends from the Carrick Hills south-westwards to Girvan, and the Dalmellington outcrop extends south-westwards from Dalmellington in a zone parallel to the Southern Upland Fault.

In the Dalmellington outcrop the Lower Devonian strata are seen to rest unconformably on folded and eroded Ordovician and Silurian strata and are themselves folded and faulted. In the Maybole outcrop the base of the sequence is not seen and the strata dip fairly gently to the north-west.

The succession consists of a lower sedimentary group overlain by lavas. The sediments at Maybole are at least 420 m thick but could be as much as 1200 m thick. They consist principally of reddish, purple or brown micaceous and feldspathic sandstones with conglomerates developed in the lower and upper parts of the exposed sequence. The lower conglomerate contains pebbles, mainly of greywacke, chert, jasper and felsitic porphyry, and the upper conglomerate has a content of chert, acid igneous rocks, sandstone and quartzite with only subordinate amounts of greywacke.

In the Dalmellington area the sedimentary group is 515 to 665 m thick. It consists of a basal conglomerate, 150 to 180 m thick, sandstones with subordinate beds of conglomerate and mudstone, 335 to 425 m thick and an upper conglomerate, 30 to 60 m thick. The conglomerates consist mainly of greywacke pebbles with chert, jasper and acid igneous rocks. The basal conglomerate at Dalmellington may be the lateral equivalent of the lower conglomerate at Maybole.

The lava sequence at Dalmellington is estimated to be about 600 m thick and in the Carrick Hills it is 300 to 450 m thick.

Upper Devonian

Upper Devonian strata were laid down unconformably on a folded and eroded surface of older rocks. Evidence for strata of Middle Devonian age is lacking and hence the unconformity represents a considerable interval of time during which there was displacement on the boundary faults and folding within the graben about NE-trending axes.

The base of the Upper Devonian as represented in the Midland Valley is marked by an unconformity and the precise position of the top of the Devonian is uncertain. This uncertainty arises because the continental facies of the Devonian, represented by the Upper Old Red Sandstone, probably continued into Carboniferous time so that the top of the Devonian occurs at an horizon within the Upper Old Red Sandstone. The upper limit of the latter division is placed arbitrarily at a locally convenient lithostratigraphical boundary, commonly at the top of a group of strata containing cornstones. This boundary is lithologically convenient but is unlikely to be laterally equivalent throughout the region. In fact interdigitation of the Upper Old Red Sandstone facies with the Cementstone Group facies of the Carboniferous is known to occur in some areas.

In an attempt to solve the problem of the boundary between the Devonian and Carboniferous, a transitional division is used on some recent Geological Survey of Scotland maps. The division is called the Devono–Carboniferous and is defined as consisting of 'strata which are of indefinite age within the so-called Upper Old Red Sandstone'.

The Upper Devonian rocks differ from those of the Lower Devonian in that they are generally finer-grained and more mature. The sediments consist mainly of fine- and medium-grained red or buff-coloured sandstones, with darker red siltstones and mudstones. Conglomerates and pebbly sandstones are less common than in the Lower Devonian and have a smaller clast-size (Plate 4.1).

There is a greater proportion of quartz grains over lithic fragments and in conglomerates there is a tendency for the proportion of vein-quartz and quartzite clasts to increase upwards in the succession. The upper part of the sequence is characterised by the presence of cornstones which are fossil soils thought to be comparable to caliche deposits (Plate 4.2). Volcanic rocks are absent.

The outcrop distribution is shown on Figure 10. In the north-west part of the area the outcrop occupies a strip on the north and west of the Clyde Plateau lavas and dips gently below them to the south-east and east. In Ayrshire the outcrops are generally aligned NW–SE and emerge from below the sedimentary cover of Carboniferous rocks on either side of the Mauchline syncline. The Upper Devonian rocks crop out in the Edinburgh area and on the north-west side of the Pentland Hills where they overstep Lower Devonian and Silurian sediments and lavas. North of the Forth the strata are known in an elongate outcrop from Kinross north-eastwards to the coast, but a parallel outcrop around the Tay estuary is poorly exposed.

Stratigraphy

The fauna contained in Upper Devonian rocks is inadequate to establish the lateral equivalence between successions in different parts of the region. Correlation is based on lithology and similar successions represent a parallel sequence of depositional environments which were not necessarily synchronous throughout the region.

The succession and correlation of sequences from the Firth of Clyde area to the Firth of Tay is shown in Figure 12.

Northern part of the Midland Valley

The thickest known sequence of Upper Devonian rocks occurs in the Firth of Clyde area where it may be as much as 3000 m thick. The earliest rocks belong to the Wemyss Bay Formation which consists of cross-bedded sandstones

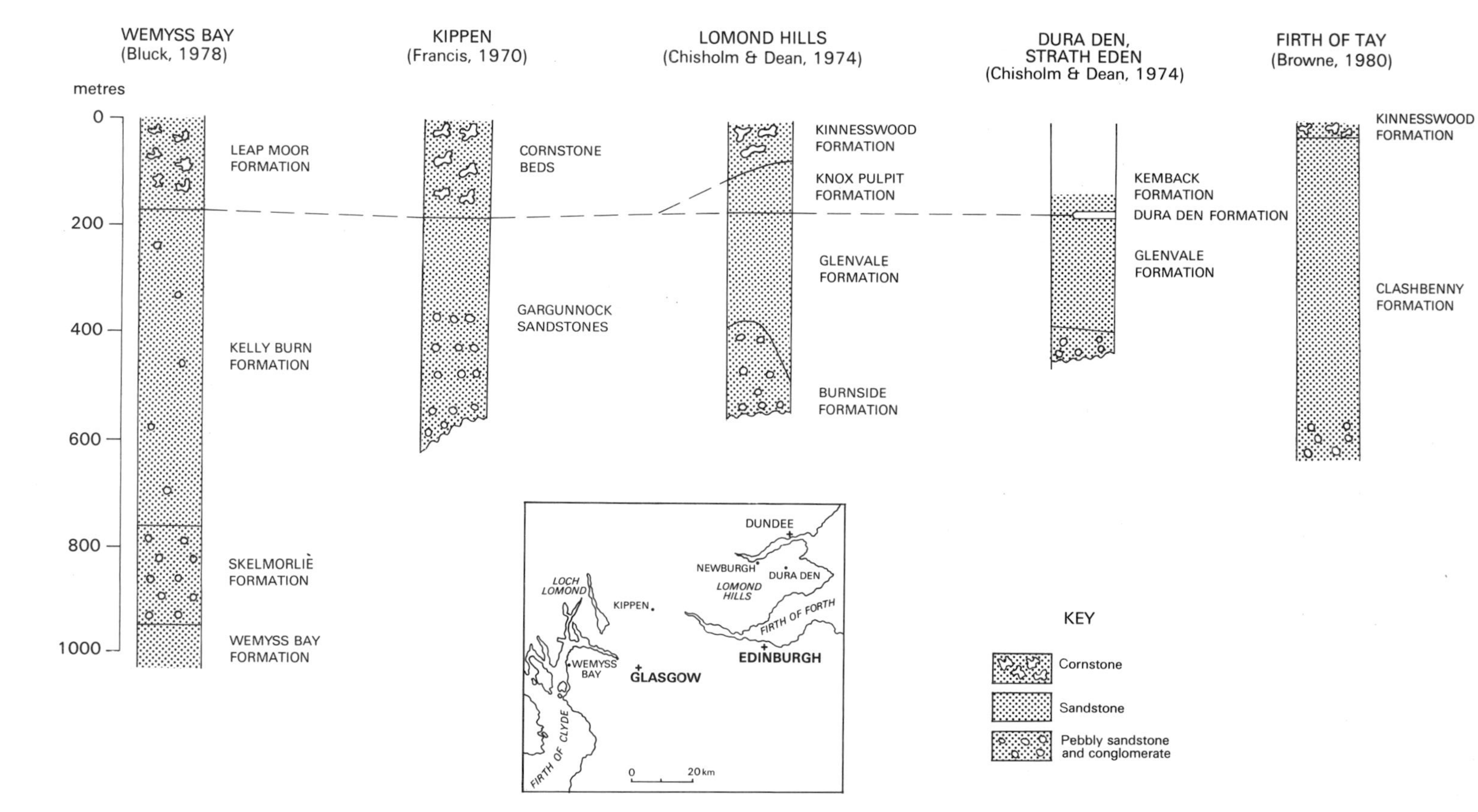

Figure 12 Correlation of Upper Devonian successions in the northern outcrops of the Midland Valley (based solely on lithological comparisons)

possibly of aeolian and fluvial overbank origin. The base of the formation is not seen.

The Wemyss Bay Formation is succeeded by the Skelmorlie Formation which is a thick conglomerate with interbedded sandstones. The conglomerate consists of clasts of metamorphic and igneous rocks and is thought to have been deposited by proximal braided streams and alluvial fans. The formation is traceable to the north-east where it is considered to be equivalent to the lower part of the Gargunnock Sandstones of the Kippen area and the Burnside Formation of the Kinross area.

The Kelly Burn Formation is predominantly sandstone and pebbly sandstone and it is equated with the upper part of the Gargunnock Sandstones in Stirlingshire which are brick-red, fine- and medium-grained sandstones with thin red siltstones and mudstones. Similar rocks are named the Glenvale Formation in Fife and Kinross and the Clashbenny Formation in Tayside. Both the Glenvale and Clashbenny Formations contain fossil fish.

Adjacent to the Highland Boundary Fault in the Firth of Clyde to Loch Lomond area a breccia and conglomerate development is interbedded with and replaces the Kelly Burn Formation. The breccias and conglomerates are distinguished from the basal conglomerate by their clast content and their palaeocurrent directions. They are thought to be alluvial fans and proximal braided-stream deposits. The sediments were derived from Dalradian and Lower Devonian rocks and were deposited as alluvial fans prograding to the south-east. This development is not represented in the successions father east.

Near Cupar, the Dura Den Formation overlies the Glenvale Formation. It has not been traced either west to the Lomond Hills area or northwards to the Tayside area. The formation is about 40 m thick and consists of red, cream and green-coloured siltstones and fine-grained, cream-coloured sandstones. The Dura Den Formation includes a well-known fish-bed containing many well preserved specimens of *Holoptychius* and *Bothriolepis*.

In the Fife and Kinross area, the Knox Pulpit Formation follows the Glenvale Formation and Dura Den Formation. It consists of soft, weakly-cemented white feldspathic sandstone and is characterised by a marked variation in grain-size between adjacent laminae. The formation differs from the other arenaceous formations in that, it cannot with certainty be described as fluviatile and there is evidence of alternating directions of current flow. The environment of deposition is possibly shallow marine indicating a marine transgression into the area from the east, but features which suggest a possible aeolian origin are also present. The formation is a good aquifer.

The youngest subdivision is predominantly sandstone characterised by the presence of cornstone. The formation has been variously named the Cornstone Beds, the Leap Moor Formation, or the Kinnesswood Formation, according to area (Figure 12). The clastic phase of the formation consists of upwards-fining cycles of sandstone, siltstone and silty mudstone. The sandstones are fine- to medium-grained, weakly cemented, and are variously coloured red, brown, yellow or white. They have an erosive base and pass up into more argillaceous beds. The sandstones are commonly from 3 to 6 m thick. The finer-grained beds consist of siltstone and silty mudstone and may be greenish or red in colour. They are usually less than 2 m thick.

The cornstones, which characterise the formation, are pedogenic carbonates (fossil soils), pale cream in colour and occurring both as concretions and in conglomerates (Plate 4.2). The concretionary cornstones are impersistent rubbly beds or layers of nodules. In conglomerates the cornstones occur as prominent clasts with smaller pebbles of vein quartz and quartzite. Concretionary cornstones occur up to 1 m thick in the north of the region and some can be thicker elsewhere.

The concretionary cornstones tend to occur at particular horizons in argillaceous beds. They commonly have an irregular base and may pass down into nodules.

The beds of cornstone are fossil soils and they represent a prolonged episode of soil-formation in semi-arid conditions during which erosion was minimal and the rate of sedimentation was low. Their occurrence has been used for stratigraphical correlation but they probably have little chronostratigraphic significance. Conditions suitable for cornstone formation also occurred in parts of the Lower Devonian and Carboniferous successions. Similar processes of soil formation occur at the present time in semi-arid climates.

South side of the Midland Valley

The Upper Devonian rocks in the southern part of the Midland Valley, from Dunbar and the Pentland hills to Ayrshire, have not received the same amount of research in recent times as the succession farther north, and they are much less well known.

The thick clastic sequences consisting mainly of conglomerates and breccias seen in the Firth of Clyde area are absent and the sequence, which is undivided, consists mainly of sandstones and mudstones with cornstones.

Basal conglomerates occur only in the Edinburgh area, in parts of the Pentland outcrop and locally in south Ayrshire.

The predominant lithology is sandstone, red or pink in colour, and less commonly purplish, yellow or white. The sandstones are fine- to medium-grained and locally calcareous. They contain thin conglomerate bands and scattered pebbles, both intra-formational and exotic. Bands of siltstone and silty mudstones are also common.

Cornstones are present throughout the sequence except in the Dunbar area where they occur only in the upper part. They are particularly common in Ayrshire where they may be up to 3 m thick and were quarried and mined at one time for agricultural lime. Locally, cornstones were developed on the surface of the underlying folded and eroded Lower Devonian rocks.

The thickness of the strata varies considerably. In the Edinburgh area the maximum is about 600 m but the rocks are apparently absent in places on the north-west side of the Pentlands and at the south-west end of the Midlothian syncline. In Ayrshire, the maximum thickness occurs near Straiton where it is estimated to be between 300 and 400 m. The strata thin to the south-west and north-west.

Upper Devonian rocks are unrepresented or are very thin in the area around Strathhaven, Lanark and Douglas.

Conditions of deposition and palaeogeography

The succession represents an upward-fining and maturing, mainly alluvial sequence. The thickest and coarsest sediments, are present in the Firth of Clyde area and they indicate vigorous erosion and rejuvenation of source areas, both within the Midland Valley and beyond it. The finer-grained sediments were deposited in the more distal parts of the alluvial system and possibly also in temporary lakes. Some features of the Knox Pulpit Formation in Fife and Kinross suggest a shallow marine environment encroaching from the east but an aeolian origin is also possible. The development and preservation of cornstones imply long periods of tectonic inactivity in a mature landscape with erosion and deposition at a minimum.

The strata were laid down in a generally eastward draining fluvial system. The earlier proximal sediments indicate progradation to the north-east but younger sediments give evidence of an easterly dipping palaeoslope. Locally, in the area of the Highland Boundary Fault, provenance from the north-west is indicated.

In Ayrshire, the main sediment supply came from the Southern Uplands, but locally also from the Carrick Hills and the area north of Muirkirk.

Palaeontology

The fossils found in the sediments consist principally of plant and fish remains. The fish faunas from Dura Den, near Cupar, and Clashbenny in Perthshire are well known and suggest a Famennian age. The fish from these localities include species of *Bothriolepis, Eusthenopteron?, Glyptopomus, Holoptychius, Phaneropleuron* and *Phyllolepis. Bothriolepis* and *Holoptychius* have also been found at various other localities including Bracken Bay, near Ayr, and near Dumbarton. The presence of *Grossolepis brandi* in the Berwickshire coast sequence, just beyond the margin of the present region, has been used to suggest the presence of Frasnian strata, but no Frasnian fossils have been recorded from the Midland Valley.

The relationships of the fish-bearing horizons to successions elsewhere are shown on Figure 13. Plant remains, none of them diagnostic have been found at various localities.

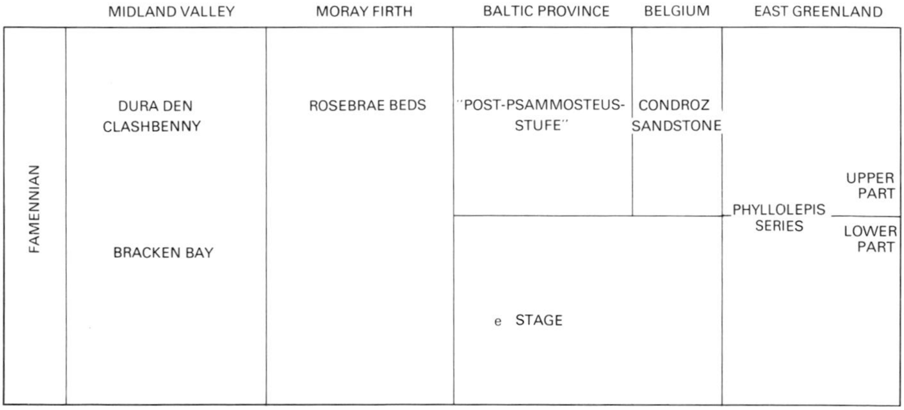

	MIDLAND VALLEY	MORAY FIRTH	BALTIC PROVINCE	BELGIUM	EAST GREENLAND	
FAMENNIAN	DURA DEN CLASHBENNY	ROSEBRAE BEDS	"POST-PSAMMOSTEUS-STUFE"	CONDROZ SANDSTONE	PHYLLOLEPIS SERIES	UPPER PART
	BRACKEN BAY		e STAGE			LOWER PART

Figure 13 Approximate correlation of the fish-bearing horizons in the Famennian of the Midland Valley (after Miles, 1963)

5. Silurian and Devonian igneous activity

A widespread Caledonian calc-alkaline igneous province extends throughout Scotland and Northern Ireland. Early Caledonian activity in the general area of the Midland Valley is indicated by the presence of igneous clasts in conglomerates of the Lower Palaeozoic inliers. The earliest penecontemporaneous volcanic deposits occur in the Downtonian, near Stonehaven and activity increased in intensity, frequency and extent to a maximum in the early Lower Devonian. Volcanicity had ended by the late Lower Devonian. Radiometric ages of Midland Valley intrusions and volcanic rocks fall within the range of 411 to 407 ± 6 Ma (Thirlwall, 1983a).

Volcanic rocks constitute a major part of most Lower Devonian successions forming such prominent uplands as the Sidlaw, Ochil, Pentland and Carrick hills (Figure 14). Most of the volcanic sequences are cut by contemporaneous dykes and thin sills. Larger, basic to intermediate hypabyssal intrusions occur in the Sidlaw Hills and in south Ayrshire and thick sills and laccoliths of acid rocks are widespread in the southern Midland Valley. Small dioritic plutonic masses with significant metamorphic aureoles occur in Ayrshire (Distinkhorn, Tincorn Hill and Fore Burn), in the western Ochil Hills and in the Pentland Hills.

Volcanic activity

Most volcanic successions consist of basaltic and andesitic lava flows with some intercalated volcaniclastic sediments ranging from siltstones to coarse conglomerates. Lava flows usually occur as impersistent, interdigitating, concordant sheets, 3 to 15 m thick. Individual flows can rarely be traced far, although distinctive groups of flows can often be correlated over considerable distances. Pyroclastic rocks are rare, except in the Pentland Hills and western Ochil Hills where they constitute a major part of the succession, and very few central vents are recognised. It thus seems likely that many of the flows emanated from fissure eruptions, the feeders to which are rarely preserved as contemporaneous, comagmatic dykes.

In some areas intercalations of fluvial and lacustrine sediments indicate that the lavas were erupted upon flood plains and that basin subsidence and erosion generally kept pace with the growth of the volcanic pile. Flow tops are often slaggy, cavernous, autobrecciated and reddened, although boles are seldom developed, suggesting that subaerial flows were rapidly covered by detritus which restricted surface weathering and laterite formation. Intercalated torrential deposits of thick, coarse-grained, volcaniclastic conglomerate indicate that considerable relief existed locally at times as volcanic hills built-up above the plains. Fine-grained, upward-coarsening, laminated sediments were deposited in lakes possibly impounded by temporary volcanic barriers.

A common feature of many areas, but particularly well seen in the coast sections of south Ayrshire and Angus are irregular intercalations and cavity-or fissure-fillings of siltstone and sandstone within the lava flows. Lamination in

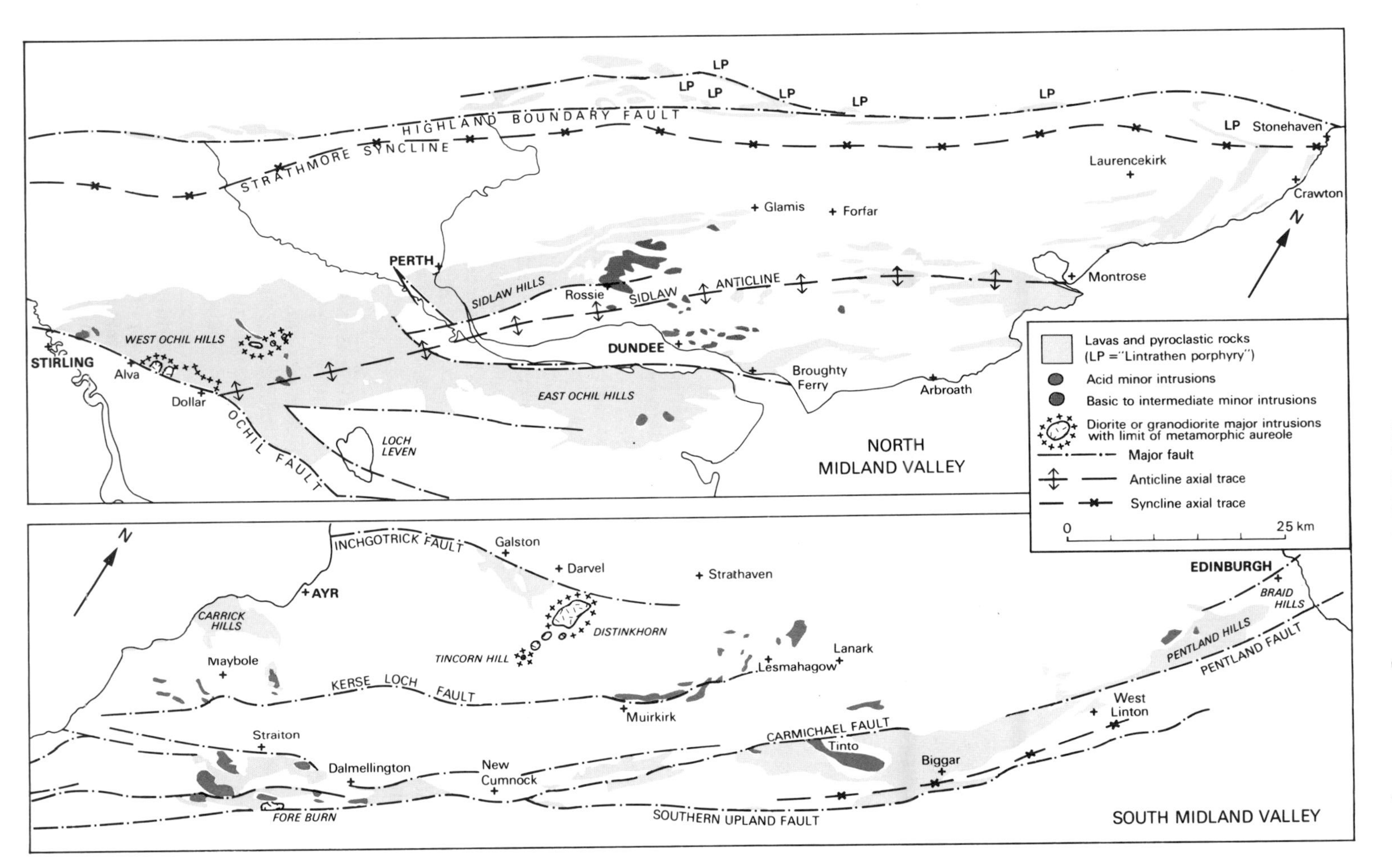

Figure 14 Outcrop of late-Silurian and Lower Devonian igneous rocks in the Midland Valley

the sediments is often undisturbed and a wide variety of sedimentary structures and trace fossils have been recorded, even from within cavities. Most authors have regarded the sediment as having been washed into voids, and in some cases lava tunnels, during rapid burial by flood deposits. However, in some cases the evidence suggests that pre-existing sediment was incorporated in later lava flows. In the Ochil Hills and on the Angus coast, features at the base of flows suggest that lavas were erupted on wet, unconsolidated sediments, possibly in a lacustrine environment. In south Ayrshire, zones of lobate pillow structures, accompanied by brecciated fragments of chilled lava (hyaloclastite) in a sandstone matrix, occur at the top and base of many andesitic sheets. These have been interpreted as sills, intruded at shallow depth into weak, unconsolidated, water-saturated sediments which became fluidised in contact with the magma (Kokelaar, 1982).

Petrology of the lavas

The extrusive rocks range in composition from olivine-basalt to rhyolite, although the dominant type in most areas is andesite or basaltic andesite. Many areas contain a complete range of compositions (e.g. Pentland Hills), but more restricted ranges occur in Ayrshire and the Sidlaw Hills where in situ acid lavas are virtually absent. Acid intrusions are common, even in areas of basic and andesitic volcanicity and it is possible that acid material may have been erupted in such areas, not as lavas but as pyroclastic deposits which were subsequently removed easily by erosion.

Basaltic and andesitic rocks are often impossible to distinguish in the field and, where altered, in thin section. This is reflected by a common grouping in some Geological Survey maps. Most of the basalts and basaltic andesites contain plagioclase phenocrysts, often with a combination of olivine (usually pseudomorphed), orthopyroxene, clinopyroxene, and rarely magnetite. Basic rocks with mafic phenocrysts alone are rare. Hypersthene-, augite- and hornblende-andesites, with or without plagioclase phenocrysts, are common and biotite-andesites occur infrequently. Trachyandesites, trachytes, dacites and rhyolites contain phenocryst combinations of oligoclase/albite, potash feldspar, quartz, hornblende and biotite. Many are flow-banded and some dacitic ignimbrites are recognised. The basic and intermediate rocks are commonly seen to have minor amounts of potash feldspar, with or without quartz, in the groundmass and sanidine rims are sometimes observed on plagioclase phenocrysts. Quartzo-feldspathic segregation veins are also common in basic and intermediate intrusions.

Amygdales are a common feature of the lavas. The siliceous varieties weather out of their matrix to form pebbles which have been collected from river gravels and from the beaches of Ayrshire and Kincardineshire to be polished as semi-precious stones ('Scotch Pebbles'). The principal decorative amygdales are agate, onyx, jasper, amethystine quartz and rock crystal but other recorded minerals include quartz, calcite, mixed carbonates, zeolites, celadonite, vermiculite and chloritic material.

North Midland Valley

Volcanic rocks and volcaniclastic sediments are present on the limbs of the Strathmore Syncline and Sidlaw Anticline. Intermittent volcanic activity commenced in the Downtonian Stonehaven Group and increased in intensity to a maximum development in the Gedinnian Arbuthnott Group. Higher groups are apparently devoid of volcanic rocks, apart from a few local flows in the Garvock Group south of Laurencekirk.

The lower parts of the Strathmore succession are exposed only in the north-east, around Stonehaven and in faulted inliers in the Highland Boundary Fault Zone. Clasts of acid lava first appear in a laterally-persistent conglomerate in the middle of the Stonehaven Group and thereafter sandstones and local conglomerates commonly contain a high proportion of angular andesitic and rhyolitic volcanic debris. True volcanic agglomerates, with fragments of pyroxene-, amphibole- and biotite-andesite occur rarely at the top of the Stonehaven Group and in the Dunottar Group. The lowest exposed lava flow, an augite-andesite, occurs in the middle of the Dunottar Group and at the top of the group, several flows of olivine-basalt constitute the Tremuda Bay Volcanic Formation. In the Crawton Group, two isolated thin flows of porphyritic andesite occur in the lower formations and at the top of the group three or more flows, separated by erosion planes and intraformational conglomerates, constitute the Crawton Volcanic Formation. This formation is characterised by distinctive macroporphyritic basaltic andesites with plagioclase phenocrysts up to 25 mm long. In the Highland Boundary Fault Zone, a persistent dacitic ignimbrite up to 40 m thick (originally described in part as the 'Lintrathen Porphyry') forms a good stratigraphic marker, probably equivalent to the top of the Crawton Group.

Volcanic rocks predominate in two of the four diachronous formations of the Arbuthnott Group:

The Montrose Volcanic Formation consists of four major volcanic developments with a collective maximum thickness of 200 m, which are thought to have emanated from a centre now covered by the North Sea (the 'Montrose Centre'). Principal rock types are olivine-basalt and feldspar-phyric pyroxene-andesite with no acid lavas and only rare pyroclastic rocks and volcanic conglomerates. Coastal sections reveal good examples of sedimentary inclusions and cavity-fillings.

The Ochil Volcanic Formation constitutes almost the entire Arbuthnott Group in the western Ochil Hills and in north Fife where it has a maximum exposed thickness in excess of 2000 m. The Broughty Ferry lavas are a thinner, north-eastern continuation of this sequence on the south-east limb of the Sidlaw Anticline. On the north-west limb of the Anticline at least 1500 m are exposed in the western Sidlaw Hills and the formation thins north-eastwards to 200 m in the eastern Sidlaw Hills where only the upper members persist as far as Glamis. Throughout the formation pyroxene-andesites and olivine-basalts predominate, although the full range of lavas includes hornblende-andesites, trachyandesites, dacites and rhyodacites. The more basic lavas are especially predominant to the north-east in the Sidlaw Hills and Broughty Ferry sequences. Intercalations of volcaniclastic conglomerates are common, particularly at the top of the sequence. Many contain a high proportion of acid igneous material. Some are widespread, forming useful local marker horizons and others occur as lenticular valley fillings and may be mudflow deposits. In situ pyroclastic rocks are confined to the western Ochils where they are thickest, coarsest and most numerous in the south, close to the Ochil Fault, suggesting that the main centre of eruption lay to the south and is now concealed beneath younger strata. Volcanic vents are not recognised within the exposed succession and it is possible that most of the flows emanated from fissure eruptions.

On the north-west limb of the Strathmore Syncline the Arbuthnott Group contains thick developments of coarse, volcanic conglomerate with subordinate fine-grained sediments and locally up to eleven lava flows, mostly of basalt or basaltic andesite.

Throughout the Lower Devonian succession conglomerates with metamorphic and/or volcanic clasts become thicker, coarser and more numerous towards the Highland Boundary Fault. In the lower groups the volcanic component is predominantly of intermediate to acid rocks not represented in the Midland Valley suggesting the presence of voluminous, contemporaneous intermediate to acid volcanicity north of the Highland Boundary Fault, either in areas presently forming the Grampians or offshore to the north-east. The dacitic ignimbrite in the Crawton Group may be a result of such activity. More-basic igneous material becomes dominant in sediments of the higher groups, presumably derived from the Midland Valley volcanic centres in the Ochil, Sidlaw and Montrose areas.

South Midland Valley

Mixed conglomerates at various horizons in Silurian inliers contain clasts of granite, fine-grained acid rocks and rarely more-basic lavas (Chapter 3). It is probable that these are derived from early Caledonian activity. In situ evidence for this is lacking apart from a thin band of white clay in the North Esk inlier which may be an altered tuff, indicating contemporaneous volcanicity in late Llandovery time.

Lower Devonian volcanic rocks occur in a series of outcrops, separated by major NE–SW faults in a zone broadly parallel to the Southern Upland Fault from the Pentland Hills to the Straiton area. Further outcrops occur to the north-west of this zone at Galston and in the Carrick Hills. Apart from the Pentland Hills, local successions are known in far less detail than those of the northern areas.

The Pentland Hills, together with the Braid Hills and Blackford Hill exhibit the thickest, most continuous and most varied sequence in the southern Midland Valley. At the northern end, up to 1800 m of lavas and interbedded pyroclastic deposits have been divided into ten groups. The main succession consists of upper and lower groups of olivine-basalts, and feldspar-phyric basaltic andesite (e.g. the distinctive 'Carnethy Porphyry') and central groups with a high proportion of andesites, dacites, trachytes, rhyolites and acid tuffs. Intercalated volcaniclastic sediments occur infrequently and the flows are considered to be entirely subaerial, probably emanating from a centre to the north where several small vents occur near Swanston and on the southern margin of the Braid Hills. To the south-west the lower groups are gradually replaced by clastic sediments and the upper groups are reduced in thickness towards Carlops where outcrops are terminated at the Pentland Fault. South-west of Carlops a few volcanic intercalations occur in Lower Devonian sediments, but these are mostly obscured by the unconformable overlying Upper Devonian or Lower Carboniferous rocks.

Further to the south-west volcanic rocks occur in a 6 to 8 km-wide zone between the Southern Upland Fault and the Pentland–Carmichael Fault System. Within this zone greywacke-conglomerates and sandstones derived from the Southern Uplands are locally well developed below the volcanic piles and, where present, clastic sediments above contain bands of coarse-grained lava conglomerate. Major outcrops occur between West Linton and Douglas (the 'Biggar Centre'); around Duneaton Water, south-west of Muirkirk; and in the Straiton–Dalmellington area. Estimates of thickness range from 600 m near Straiton to 1200 m near Muirkirk. North-east of the River Clyde the sequence includes many flows and a few pyroclastic deposits of acid material (trachyte, rhyolite and brecciated rhyolitic tuff), similar to those of the Pentland Hills, in addition to predominant olivine-basalts, pyroxene-andesites and hornblende-andesites. South-west of the River Clyde the sequences consist almost entirely of olivine-basalt and basaltic-andesite, although acid rocks are present in several hypabyssal intrusions, including the large laccolith of Tinto (p.43). Intermediate and acid lavas also occur as pebbles in sedimentary intercalations of the Dalmellington area. Pyroclastic deposits are generally rare, except locally at the base of the lava succession, and only one small vent has been recognised, at Little Shalloch, Dalmellington.

In the Carrick Hills, south-west of Ayr, a volcanic pile of 300 to 450 m rests upon a thick sequence of clastic sediments and is overlain unconformably by Upper Devonian sediments. Related volcanic rocks occur south of Dalrymple and at Maidens. The sequence consists almost entirely of olivine-basalts and hypersthene-andesites and has several features in common with that of the Dalmellington area with which it may be contemporaneous. Coastal sections exhibit a variety of sedimentary inclusions, infillings and intercalations (p.36). Pyroclastic deposits are rare, but possible contemporaneous vents occur at Mochrum Hill and Bracken Bay.

Poorly-exposed high ground to the south of Galston and Darvel is composed of feldspar-phyric olivine-basalts and hypersthene-andesites with numerous intercalations of volcaniclastic conglomerate and sandstone. Decorative jasper and agate has been quarried and collected as stream pebbles, particularly in the Burn Anne area.

Minor intrusions

Dykes and thin sills

The majority of minor intrusions exposed in the Midland Valley are closely associated with sequences of Lower Devonian volcanic rocks, where dykes and thin sills are widespread and ubiquitous but rarely abundant. They are seen to cut the volcanic sequence and the underlying sedimentary successions in all areas but are never found in overlying sediments, and are thus regarded as contemporaneous with the volcanic activity. Outside of the areas of known volcanicity, dykes are numerous around the Distinkhorn granodiorite-diorite complex where they cut Lower Devonian sediments and are thermally metamorphosed by the pluton. A few dykes within the aureole have not been metamorphosed and represent a slightly later generation. The only dykes not associated with any known igneous centre cut Lower Devonian and more rarely Silurian sediments of the Lesmahagow Inlier.

The dykes do not exhibit a constant regional trend, although there is a tendency for them to occur in the NE–SW quadrant as in the Pentland Hills. A swarm of dykes in the western Ochils is radially disposed around a group of plutonic intrusions (p. 46). The majority of dykes are 1 to 2 m in width, although widths up to 6 m are common and 25 m-wide dykes are recorded in the Ochils. Most are impersistent laterally or cannot be traced beneath drift cover, but distinctive dykes have been traced for up to 3 km in the Ochil and Pentland Hills.

The majority of dykes are of basic to andesitic composition with close petrographic and geochemical similarities to the lavas. Rock types include olivine-dolerite or basalt, quartz-dolerite, porphyritic microdiorite or andesite (with combinations of plagioclase, orthopyroxene, clinopyroxene, hornblende and biotite phenocrysts), microgranodiorite (with or without quartz and/or albite phenocrysts) and microgranite. Many are highly altered by albitisation, sericitisation, chloritisation, carbonation and silicification such that all traces of primary minerals are obliterated and classification, where possible, is based upon recognition of pseudomorphs and relict textures. The classification employed on Geological Survey maps reflects these difficulties and includes several general terms (Table 3). Distinctions between several of these groups are ill-defined.

Table 3 Classification of the more common rock-types of Lower Devonian minor intrusions as used in Geological Survey maps

Name on map	Compositional range	Remarks
Sub-basic and basic, mafic and semi-mafic	Olivine-dolerite/basalt quartz-dolerite	Rarely porphyritic with labradorite or andesine phenocrysts. Commonly severely altered
Microdiorite		Non-porphyritic
Porphyrite and basic porphyrite where differentiated	Hypersthene-, augite- or hornblende-andesite/microdiorite	Porphyritic. Abundant phenocrysts of albitised plagioclase with or without pyroxenes and/or amphibole
Acid porphyrite	Hornblende- or biotite-andesite, trachy-andesite, dacite, microgranodiorite	Porphyritic. Abundant phenocrysts of albitised plagioclase with or without hornblende or biotite, rarely pyroxene. Sparse quartz and/or alkali feldspar in groundmass. More leucocratic than porphyrite.
Quartz-albite-porphyry and quartz-porphyry	Microgranodiorite/rhyodacite, micro-granite/rhyolite	Porphyritic. Phenocrysts of quartz with albite (primary or albitised) and/or potash feldspar. Biotite usually present.
Felsite	Microgranodiorite/rhyodacite, micro-granite/rhyolite	Sparsely porphyritic or non-porphyritic, fine-grained, often devitrified, compact leucocratic rocks. Usually pink-weathering
Plagiophyre	Mostly andesitic	A locally convenient, vague term for highly altered, sparsely porphyritic to non-porphyritic, rocks with albitised feldspar.
Lamprophyre	Mostly kersantite	Essential plagioclase and biotite

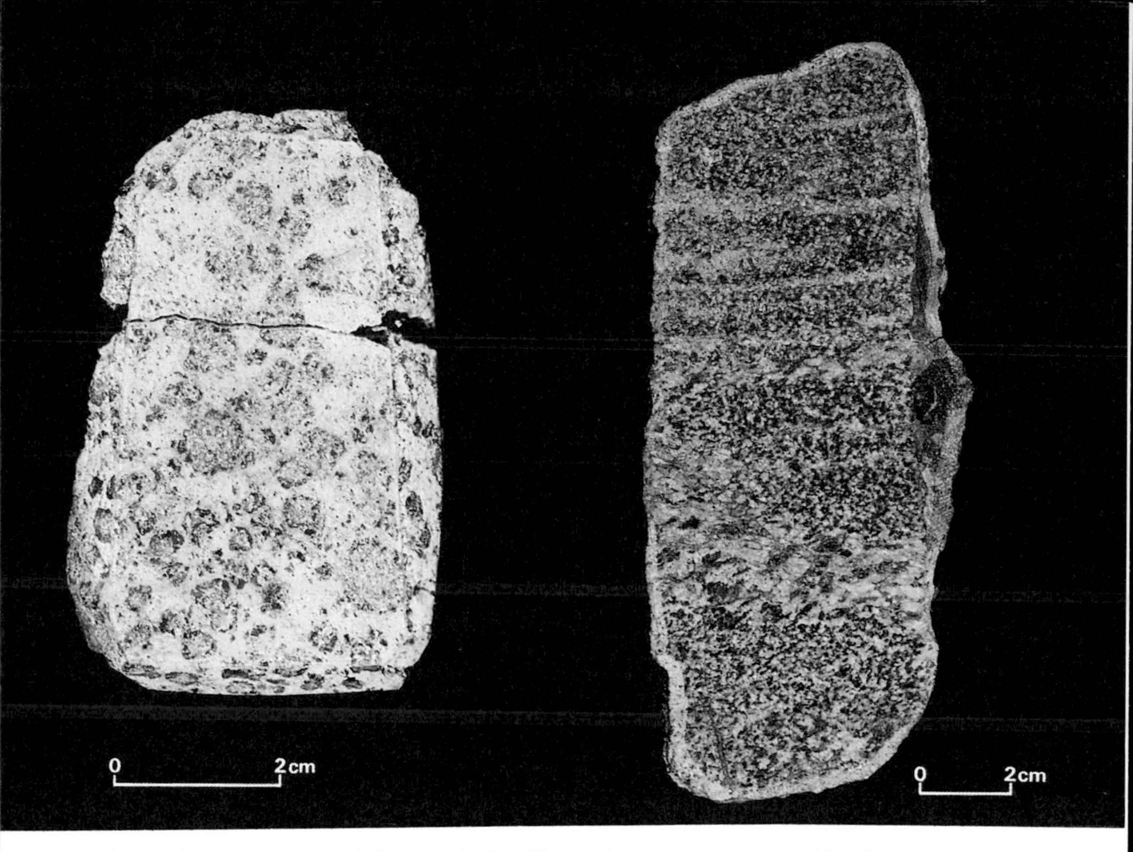

1 Inclusions of basement material from Carboniferous Partan Craig vent, North Berwick (*Grant Institute of Geology, University of Edinburgh collection*): Left, garnet-rich, quartzo-feldspathic granulite; Right, banded pyroxene granulite. (MNS 3670, 3671)

Plate 3

2 Ochil Fault scarp near Stirling. Lower Devonian lavas of Ochil Hills faulted against Carboniferous sediments under plain of Carse of Stirling. Stirling Castle and Wallace Monument perched on quartz-dolerite intrusions. (D 1941)

1 Upper Devonian sandstone and conglomerate near Wemyss Bay, Firth of Clyde coast. (D 1557)

2 Cornstone and red sandstones, Inverkip, Firth of Clyde coast. Several fossil soil horizons in sandstone, with vertically disposed concretionary nodules. (D 2596)

Plate 4

Intrusions of the Dundee area

Numerous thick sills, laccoliths and bosses cut the Ochil Volcanic Formation of the Sidlaw Hills and the underlying sediments of the Dundee Formation in the core of the Sidlaw Anticline. The largest intrusion is a 100 m-thick sill extending for 7 km from Rossie Hill to Lundie. Other sills and laccoliths can be traced for up to 2 km and bosses averaging 0.5 km in diameter are widespread. An undulating laccolith roof, exposed in the railway cutting east of Ninewells has produced conformable dome and basin structures in overlying sandstones and mudstones.

The intrusions are probably contemporaneous with or immediately post-date the lavas, which they resemble in petrography and geochemistry. Coarse-grained quartz-hypersthene-microdiorite or dolerite ('basic porphyrite') is the most abundant rock type, including the Rossie Hill–Lundie sill and many of the bosses. Ophitic and microporphyritic olivine-dolerite occurs in irregular sill-like bodies in the Sidlaw Hills and in small bosses in Dundee. Porphyritic augite-andesites and hypersthene-andesites occur around Dundee (e.g. Dundee Law and Craigie Hill) and more-acid porphyritic rocks include the biotite-dacite of the Ninewells railway cutting. Other acidic intrusions include albitised porphyritic quartz-trachytes in small sills to the north of Dundee and in more persistent sills along the length of the Sidlaw Hills. Many of the more basic intrusions are cut by segregation veins and diffuse patches of quartzo-feldspathic material.

'Felsite' sills and laccoliths of the Midland Valley

Thick sills and laccolithic intrusions, composed of a variety of fine-grained acid igneous rocks and collectively referred to as 'felsite' and 'acid porphyrite' are abundant in the south of the Midland Valley where they are particularly concentrated in a 6 to 12 km-wide zone on the north-west side of the Southern Upland Fault (Figure 14). In the north they are represented by rare dykes and by larger intrusions in the Ochil Hills (e.g. Forret Hill and Lucklaw Hill). The 'felsites' are usually resistant to erosion relative to surrounding rocks and commonly form prominent, rounded hills such as Tinto (near Biggar), Garleffin Fell and Glenalla Fell (near Straiton). They have been extensively quarried for road metal and are a major source of the pink 'felsite chips' characteristic of so many Scottish roads.

The intrusions cut Silurian to Lower Devonian sediments below the Lower Devonian volcanics and the volcanics themselves with little thermal effect. In places they are overlain unconformably by the Upper Devonian or Lower Carboniferous and contribute distinctive pebbles to basal conglomerates. Sills are concordant with the strata or are gently cross-cutting with several leaves ranging in total thickness from 20 to 500 m. Larger laccolithic intrusions, characterised by transgressive upper margins, include the mass of Tinto which crops out over 10 km^2 and has an estimated maximum thickness of 1000 m.

Rock-types range from non-porphyritic microgranodiorite to porphyritic dacite, rhyodacite (quartz-albite-porphyry and quartz-porphyry) and rhyolites. Biotite microphenocrysts are common and pyroxene and hornblende are found in some sheets. Garnet has been reported from the micro-granodiorite of Tinto.

Intrusions of the Maybole–Straiton–Dalmellington area

A group of thick concordant sills of basic to intermediate composition intrude Lower Devonian sediments below the volcanic sequence, to the south and west of Maybole. Similar rock types occur in sills and dyke-like bodies intruded at similar stratigraphic levels to the south of Straiton and Dalmellington. The intrusions consist mainly of highly-altered andesitic rocks ('plagiophyres') and quartz-dolerites. The latter contain interstitial quartz but no micropegmatite in contrast to the late-Carboniferous quartz-dolerites of the Midland Valley (p.120). Decomposed quartz-free dolerites are less common but there are a few olivine-basalts some of which resemble local lava types. A few lamprophyre sills of limited lateral extent are mostly classed as kersantites.

Major intrusions

Several of the larger intrusions of the Midland Valley have plutonic characteristics and are surrounded by wide thermal metamorphic aureoles.

Distinkhorn and Tincorn Hill granodiorite-diorite complexes

A group of poorly-exposed intrusions to the south-east of Darvel constitutes the only major representative of the Caledonian granodioritic plutonic suite in the Midland Valley (Figure 15). The intrusions are small in comparison with those of the Highlands and Southern Uplands but rock types and field relationships are comparable. Three separate outcrops occur within a metamorphic aureole which extends NNE–SSW for 10 km in Silurian and Lower Devonian sandstones. Contacts with the sediments are not exposed but the width of the aureole (up to 1 km in places), relative to the size of the intrusion, suggests that they may dip outwards at a shallow angle.

The Distinkhorn and Glen Garr outcrops consist of pink or red hornblende-biotite-granodiorites and grey quartz-diorites or hypersthene-diorites. The dioritic rocks occur on the eastern margins and have undergone contact metamorphism, presumably by the granodiorite. Contact-altered, fine-grained dioritic rocks also occur in a 1 km-wide intrusion on the east side of the aureole at Hart Hill. The Tincorn Hill complex consists of a variety of granodioritic and dioritic rocks, surrounded by a separate, 1 km-wide aureole in Lower Devonian sediments. A vent of agglomerate and dolerite at Auchinlongford may be a related feature.

Numerous basic to acid dykes and small sills of Lower Devonian age which cut baked sandstones in the aureole of the Distinkhorn mass have been subjected to varying degrees of contact metamorphism. Progressive recrystallisation in the dykes is accompanied by the development of epidote, chlorite, hornblende, biotite, augite and sphene with clouding of feldspar. The same phases are found in lesser amounts in the recrystallised sandstones and some tourmaline has been recorded.

Fore Burn dioritic complex

The complex occupies an area of 2 km by 0.5 km within Lower Devonian andesites, basalts and sediments adjoining the Southern Upland Fault, 7 km south-east of Straiton. Varieties of quartz-diorite are predominant with younger intrusions of albitised porphyritic andesite. Adjacent sediments are highly baked and the lavas are epidotised. All the rocks of the complex and to

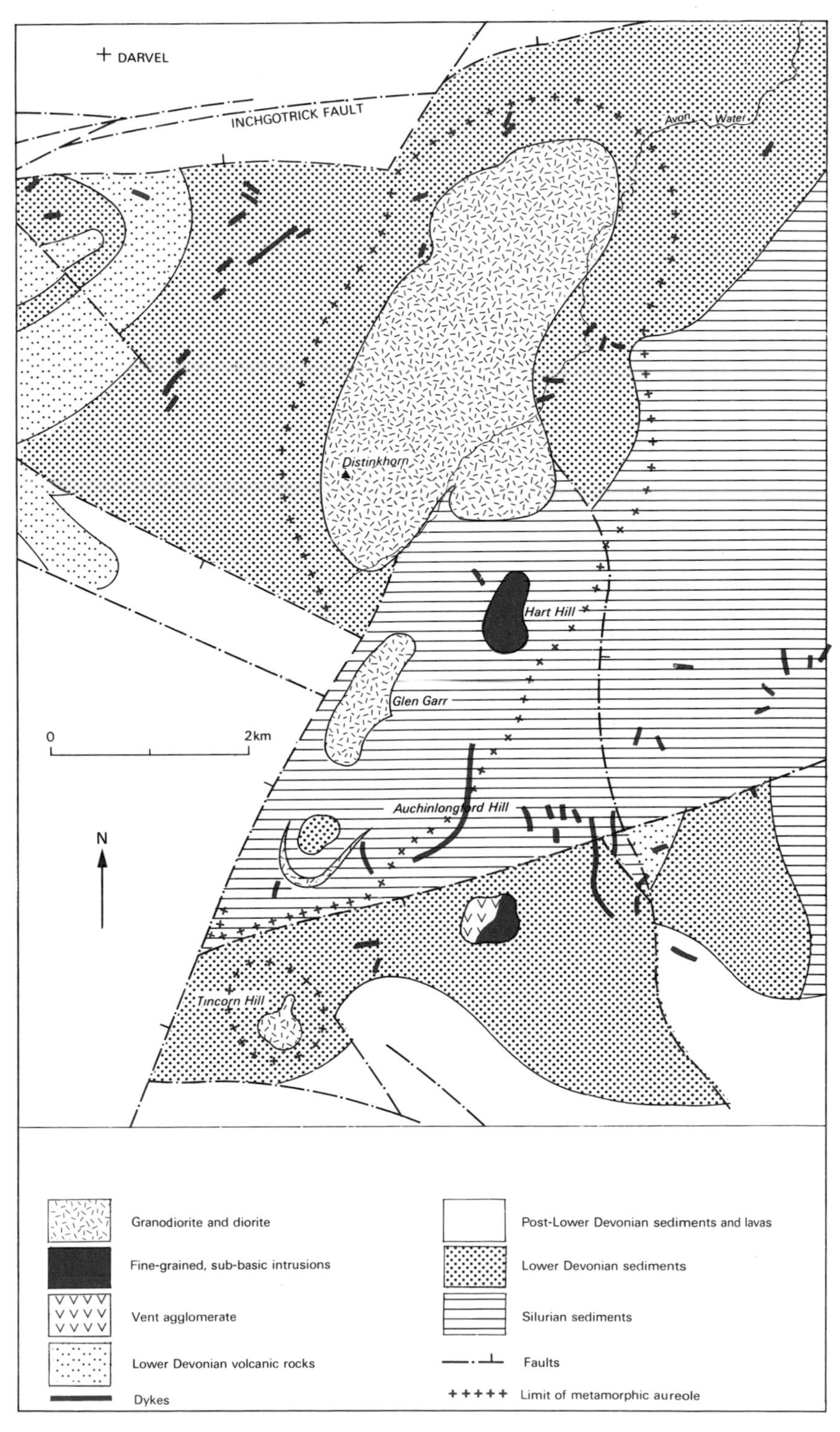

Figure 15 Lower Devonian igneous rocks in the area around Distinkhorn and Tincorn Hill, Ayrshire

a lesser extent the contact-altered lavas and sediments are affected by tourmalinisation which is locally very intense. Apatite is commonly associated with the tourmaline and thin sulphide veins contain low-grade copper-arsenic-antimony mineralisation.

Lyne Water dioritic intrusion

An outcrop of dioritic rocks cuts and bakes Lower Devonian conglomerates south-west of Wether Law in the Pentland Hills. Rocks of the intrusion are similar to those of other small masses regarded as Caledonian in the north-eastern Southern Uplands. The predominant rock type exposed is quartz-diorite becoming more basic and finer-grained south-eastwards towards a marginal porphyritic augite-andesite. In the north-west, granodiorite and microgranite outcrops suggest that a more extensive, granodioritic central part of the intrusion may be present beneath the unconformable Upper Devonian cover.

Dioritic stocks of the Ochil Hills

Dioritic stocks cut Lower Devonian volcanic rocks at Tillicoultry, Glendevon and near Glenfarg.

At Tillicoultry four stocks, ranging from 200 m to 1 km diameter, occur within a 6 km by 1 km thermal aureole. The stocks are cut by the Ochil Fault, by the late-Carboniferous quartz-dolerite fault intrusion and by members of a contemporaneous radial dyke swarm. At Glendevon a 250 m by 1 km stock and three small bosses occur within a 4 km by 1.5 km thermal aureole. Two small stocks occur at Newhill, west of Glenfarg and widespread areas of alteration elsewhere could indicate concealed intrusions. Some stocks have sharp, steep, outward-dipping intrusive margins, some have gradational, hybridised xenolithic contacts and others exhibit a ghost stratigraphy with scarp features conformable with the regional dip of the lavas. It may be concluded that the diorites were emplaced partly by simple intrusion and assimilation with radial fracturing of surrounding rocks, and partly by metasomatic replacement of country rock.

Rocks of the major intrusions are all classed as quartz-diorites but they are variable in colour, texture and mineralogy. Many of the radial dykes are similar in composition to the lavas, consisting of olivine-basalts, porphyritic pyroxene- and hornblende-andesites, trachyandesites and albitised equivalents. Within the diorites the dykes usually have irregular, gradational and unchilled contacts indicating near-contemporaneity. Many veins of pink aplite permeate the stocks and the altered rocks of the aureoles. Within the aureoles, altered lavas show a complete gradation from clouded and spotted rocks to hornfelses in which original texture has been obliterated and replaced by granoblastic aggregates of sodic plagioclase, quartz and biotite, often with pink porphyroblasts of alkali feldspar.

Magma genesis and tectonic setting

The lavas of the Midland Valley, together with contemporaneous volcanic sequences in the Grampian Highlands, constitute a widespread province characterised by many petrological and geochemical similarities. In several areas many of the more basic lavas contain high levels of such elements as magnesium, nickel and chromium which suggests that they represent primitive magmas originating by partial melting of upper mantle material. However, relatively high levels of incompatible trace elements, present in only small amounts in mantle material, suggest that the melts may have been enhanced by material from some other source. A recent isotope and rare-earth element study suggests that altered basalts and/or sediments from oceanic crust may have provided such a source.

Lava suites throughout the province are of a calc-alkaline nature and many are of a relatively high-potash type with a trace-element chemistry characteristic of active, orogenic continental margins. Spatial variations in trace element concentrations and ratios, in certain isotope ratios and in some major elements have been recognised in a NW–SE direction, perpendicular to the Caledonian structural trends. Differences are most marked between the Grampian Highlands and north Midland Valley, and slight changes across the Midland Valley are in the same sense. The relationships suggest a model of magma genesis involving melting of mantle material above a descending slab of oceanic lithosphere in a NW-dipping subduction zone (Thirlwall, 1981, 1982).

This model is broadly consistent with most current theories of Caledonian plate tectonics. Such theories invoke the closure during Silurian and Lower Devonian time of an ocean (the Iapetus Ocean), which separated an Anglo-European continent from North America and Scotland throughout most of lower Palaeozoic time. On the north-western side of this ocean, trench and ocean floor sediments were accreted on to the continental margin, deformed and uplifted in a subduction complex which is now exposed in the Southern Uplands. To the north-west of this complex, on the site of the future Midland Valley, Silurian marine sediments were deposited in a shallow forearc basin, which later became filled with Lower Devonian fluvial sediments as subduction-related volcanic and plutonic activity became widespread. The cessation of igneous activity during the later Lower Devonian is taken to indicate that continental collision and suturing had occurred and that subduction had ceased. Subsequent tensional release then created or rejuvenated major crustal fractures and deep molasse basins formed in an incipient Midland Valley graben which was eventually to develop into the continental rift in which the alkaline volcanics of the Lower Carboniferous were erupted.

6. Carboniferous: General

Rocks of Carboniferous age underlie the major part of the Midland Valley. The subdivisions and classification of the sequence are shown on Table 4. Descriptions of the sediments are given in the Dinantian, Namurian and Westphalian chapters and the palaeontology and igneous rocks are also treated in individual chapters.

The presence of Carboniferous rocks in central Scotland played a fundamental part in the industrial development of the region. The location of the iron and steel industry and the consequent developments in heavy engineering were in large part due to the existence of seams of coal, ironstone and limestone in the Carboniferous rocks. Although ironstone mining has long since finished, coal-mining remains an important, though shrinking industry. The mining, quarrying and sinking of boreholes associated with the exploration and exploitation of these minerals has resulted in a vast amount of detailed knowledge of the Carboniferous rocks in the region, particularly in the coalfield areas.

The basal Carboniferous strata in the region show a transition from the red sediments of the Upper Devonian deposited in fluviatile and lacustrine regimes to the predominantly grey, fluvio-deltaic and shallow-marine beds of the Carboniferous. This change of facies reflects a major change of climate and

Table 4 Subdivision and classification of the Carboniferous succession in the Midland Valley

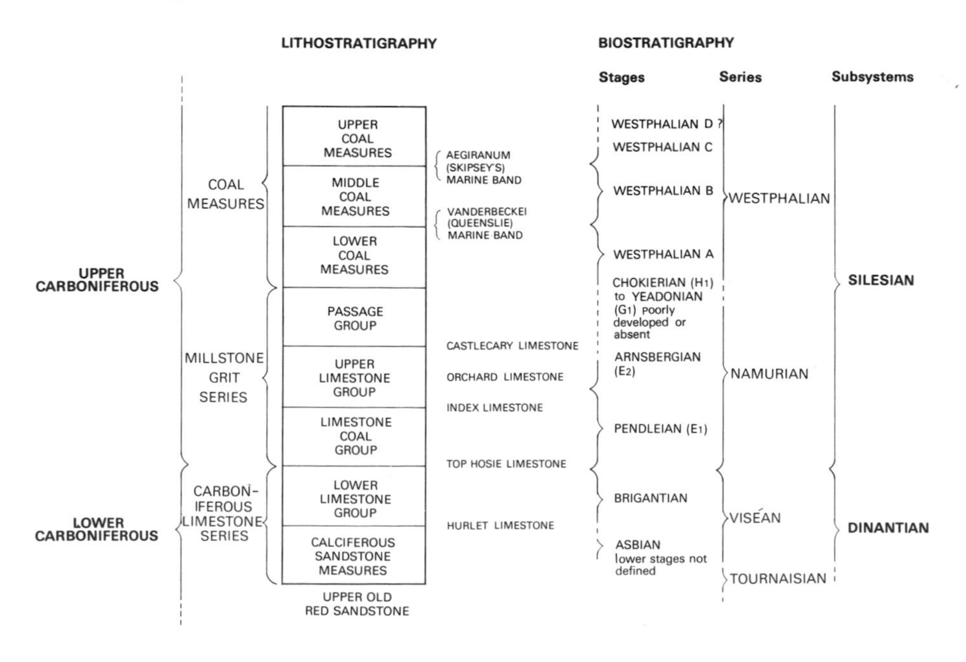

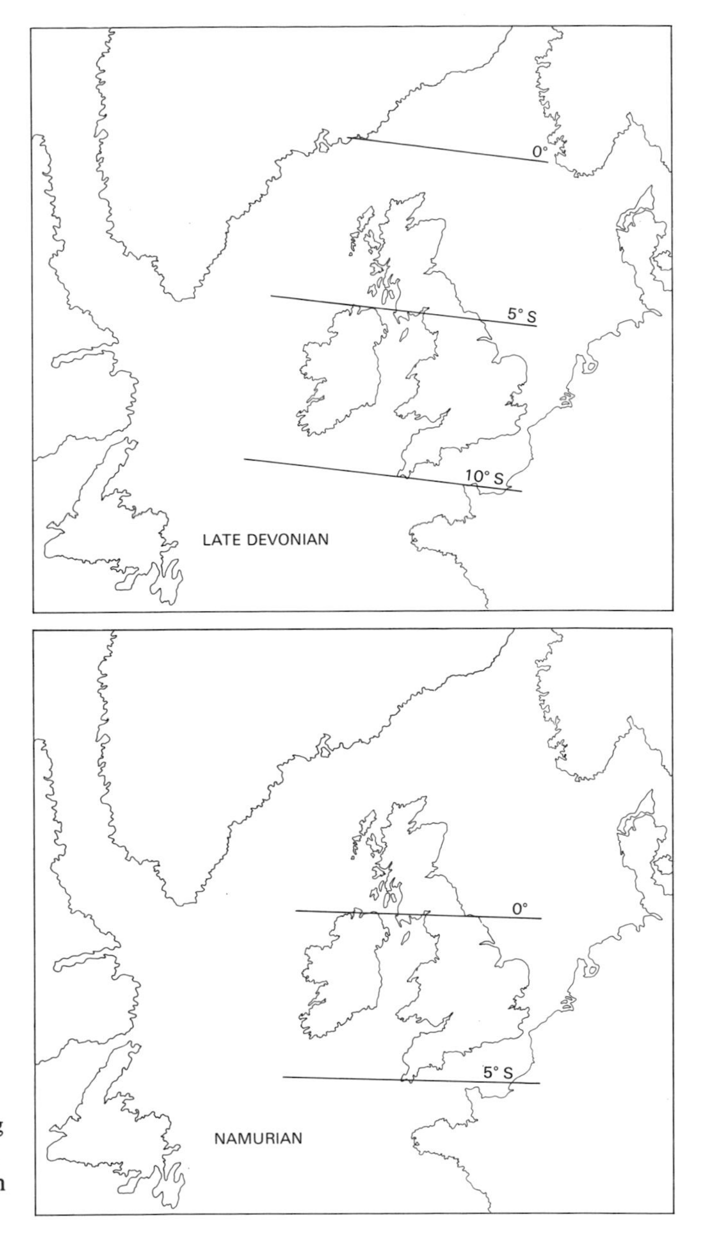

Figure 16
Palaeolatitudes for the British Isles during the late-Devonian and Namurian (after Smith and others, 1981)

depositional environment. The terrestrial, arid or semi-arid conditions during the Devonian gave way to a humid environment in the Carboniferous with deposition in a fluvio-deltaic system adjacent to and periodically transgressed by the sea. The formation of the numerous coal seams in the succession resulted from luxuriant plant growth suggesting wet and warm or tropical conditions. Palaeomagnetic evidence indicates a northward drift of the British Isles across equatorial latitudes during the Carboniferous period (Figure 16).

The red coloration of the Upper Coal Measures, the youngest Carboniferous sediments preserved in the region, points to a return to semi-arid conditions towards the end of the period.

In Carboniferous times the Midland Valley lay near the northern edge of a platform on the south side of the North American–North European continent. Clastic sediments derived from the Caledonian Mountains in the north interdigitated with thin marine sediments deposited during periodic incursions by the sea. The acme of marine conditions occurred at the close of Dinantian times during the deposition of the Lower Limestone Group and marine sediments are also well developed in the Upper Limestone Group in the Namurian. Repeated movements on deep fractures caused differential subsidence over the region which led to lateral variation in thickness and facies of the sediments.

The area now occupied by the Southern Uplands was probably above sea level for most of the Carboniferous except for gaps in the Loch Ryan and Sanquhar–Thornhill areas and possibly on the Berwickshire coast. Within the Midland Valley the thick subaerial accumulation of the Clyde Plateau volcanic rocks formed a barrier to sedimentation in Lower Carboniferous times and this barrier was not completely submerged until late in the Dinantian.

The Carboniferous sediments consist mainly of sandstones, siltstones and mudstones with beds of limestone, coal, fireclay and lesser amounts of ironstone and oil-shale. They are all shallow-water deposits and despite the great variation in total thickness in different areas, the depositional surface was never far from sea level and subsidence and sedimentation more-or-less kept pace.

Throughout much of the succession the sediments are arranged in cyclic sequences with a complete cycle represented by an upward passage from marine limestone or mudstone through non-marine mudstones and sandstones to seat-bed and coal, followed by a return to a marine member. This cyclicity is usually imperfect with one or more members of the cycle missing. The cycles can be up to 30 m thick but the average thickness is about 10 m. The marine members of the cycles are of primary importance in the correlation of sequences from different areas. In general, the number of cycles in a succession increases in direct proportion with increase in thickness and more cycles tend to be complete in thicker successions.

Several explanations for the cyclicity exhibited in the deposition of Carboniferous sediments have been proposed. These include eustatic changes in sea level, pulsed subsidence and purely sedimentary processes inherent in delta construction. No single explanation finds universal acceptance and it is likely that several mechanisms were interacting during the Carboniferous.

In international classification the base of the Carboniferous, i.e. the base of the Dinantian (Lower Carboniferous), is defined by the presence of certain marine fossils. As these fossils have not been found in Scotland, an accurate definition of the base of the Carboniferous in the region cannot be drawn. Plant miospores are the only fossils present in the lowest Carboniferous beds in the region which give some indication as to where the horizon should be placed. These minute fossils have been used to subdivide the Dinantian into a number of assemblage zones. In this zonation, the Cementstone Group, the lowest undoubted Carboniferous strata in the Midland Valley are placed in the CM Zone. In southern England and elsewhere, however, there are at least two lower zones present below the CM Zone. As the Cementstones overlie the Upper Old Red Sandstone in apparent conformable sequence, there is a strong

1 Fossil Grove, Victoria Park, Glasgow. Natural casts of Carboniferous (Namurian) trees. (D 1536)

2 Top of Dockra Limestone, Lower Limestone Group, Old Mill Quarry, Beith. Top surface is glaciated and has reddened crust on pale rooty top bed. (D 2762)

Plate 5

1 Lower Carboniferous bedded trachytic agglomerates and tuffs with large ejected lava blocks, Ardoch Burn, Carrot, Eaglesham. Bedding distorted around blocks by impact and later differential compaction. (D 1574)

Plate 6

2 Dumbarton Rock. Plug of microporphyritic olivine-basalt of Lower Carboniferous age. (D 3494)

inference that the Devonian–Carboniferous boundary lies at an horizon within the Upper Old Red Sandstone.

The top of the Dinantian Subsystem, and therefore the base of the Silesian Subsystem, is drawn about a metre below the Top Hosie Limestone where the lowest occurrence of the goniatite *Cravenoceras* has been recorded.

The Dinantian is subdivided into two series, the Tournaisian and the Viséan. The presence of the former has been deduced solely from miospore evidence and is not well defined. The bulk of the Lower Carboniferous rocks of the region are of Viséan age.

The overlying Silesian strata present in the region are subdivided into the Namurian and Westphalian series which in turn are composed of stages as shown in Table 4. The base of the Westphalian is defined in Western Europe by the presence of the goniatite *Gastrioceras subcrenatum*. This species has not been found in Scotland and spore evidence suggests that the base of the Westphalian lies within the upper part of the Passage Group. This lack of faunal control has resulted in the base of the Coal Measures being drawn at locally convenient horizons in the various coalfield areas. Deposits of the youngest stage of the Silesian in Europe, the Stephanian, have not been recognised in Scotland.

Some of the Namurian and Westphalian stages are defined by the presence of diagnostic goniatites but for others the evidence is lacking. For descriptive purposes it is more convenient to use the classification based on lithology shown on the left hand side of Table 4.

The Dinantian succession is divided into two parts. The older one, the Calciferous Sandstone Measures, is much more variable in lithology and is much thicker than the overlying Lower Limestone Group (Figure 17).

The rocks present include the normal Carboniferous sediments of the region (p.50) but the Calciferous Sandstone Measures are noted for a thick development of cementstones at the base in many areas, oil-shales and non-marine limestones in the Lothians and parts of Fife. The sequence is interrupted in parts of the region by volcanic rocks. It is only at the top of the succession that richly fossiliferous marine strata are developed. Marine limestones and mudstones continue throughout the Lower Limestone Group but sandstones and coals are also present especially in those areas where there is a thicker development of the group.

The basal cementstone facies is interpreted as having been deposited in a lagoonal, coastal-flat environment under conditions of high salinity and periodic desiccation. It was succeeded by a fluvial and deltaic regime with the

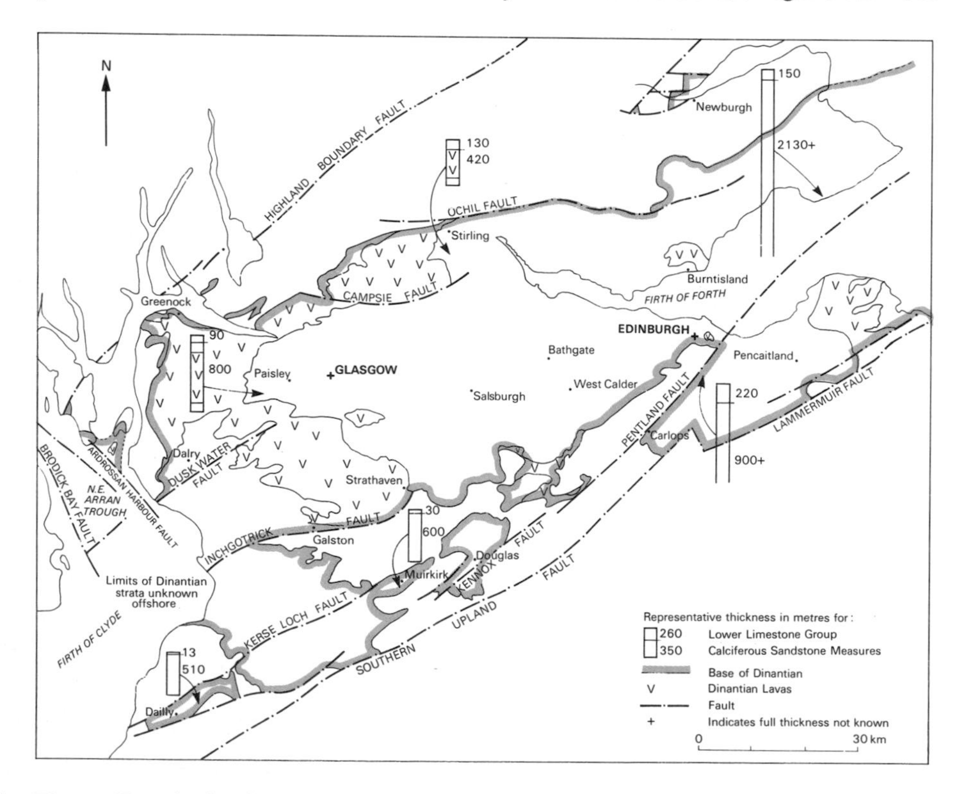

Figure 17 Distribution of Dinantian rocks in the Midland Valley. Offshore information is based on geophysical data and some boreholes

Plate 7 Carboniferous marine fossils

1 *Sanguinolites costellatus.* **2** *Antiquatonia costata.* **3** *Semiplanus* cf. *latissimus.* **4** *Schizodus pentlandicus*, ×1.5. **5** *Tornquistia youngi*, ×6. **6** *Bellerophon randerstonensis.* **7** *Liralingua wilsoni*, ×2.5. **8** *Rugosochonetes? skipseyi*, ×2.

Plate 8 Carboniferous non-marine fossils

1 *Naiadites obesus*, ×2. **2** *Curvirimula candela*, ×2. **3** *Tealliocaris woodwardi*, ×1.5. **4** *Anthracosia atra*. **5** *Paracarbonicola sp.*, ×1.5. **6** *Anthraconaia williamsoni*. **7** *Carbonicola pseudorobusta*. **8** *Telangium affine*. 9 *Sphenopteris cymbiformis*.

source of the sediments to the north. Some marine transgressions occurred in the eastern part of the region. Towards the close of Dinantian times the marine incursions affected most of the Midland Valley.

The sedimentary sequence was interrupted by volcanic activity in several areas. The large pile of Clyde Plateau lavas in the western part of the region formed an emergent land-mass which subdivided the area of sedimentation. This pile of lavas helped to form the restricted depositional environment in which the West Lothian oil-shales were laid down. Organic matter was able to accumulate during periods of minimal influx of sediment and the lack of disturbance ensured the anaerobic conditions required for the accumulation of kerogen-rich shales.

The sediments are for the most part poorly dated and correlation within the region is tentative until late in the Dinantian.

Calciferous Sandstone Measures

The Calciferous Sandstone Measures is the lower and thicker of the two divisions of the Dinantian in the Midland Valley. It succeeds the Upper Devonian rocks apparently with conformity in most areas.

As stated previously (p.50), the position of the boundary between Devonian and Carboniferous rocks cannot be defined in the region. A major facies change occurs, probably in the basal part of the Carboniferous and the rocks involved do not contain diagnostic fossils. The Upper Old Red Sandstone facies passes up into the cementstone facies of the Carboniferous and locally the two facies alternate. The base of the Carboniferous has been taken arbitrarily at the base of the lowest group of cementstones and this is unlikely to be synchronous throughout the area.

The Calciferous Sandstone Measures is the most variable of the subdivisions of the Carboniferous both in terms of its range of lithologies and in its thickness. The lithology varies from fluvial sandstones and lagoonal cementstones to marine limestones and shales, and includes non-marine limestones, oil-shales and thick volcanic formations. There is a striking contrast between the succession in the west where there are extensive thick piles of lava and the succession in the east where there is a thick sedimentary sequence (Figure 18). The thickness in east Fife is known to exceed 2 km.

The top of the Calciferous Sandstone Measures is at the base of the Hurlet Limestone which can be recognised in most parts of the Midland Valley. The sequence is subdivided on the basis of local lithology, usually into three parts. The limits of the subdivisions are not defined palaeontologically and are unlikely to be laterally equivalent

West-central area

In the west-central part of the Midland Valley the sequence is dominated by the thick lava piles of the Renfrewshire and Kilpatrick hills and the Campsie Fells. The sequence has been subdivided into a Cementstone Group underlying the lavas, the lavas themselves and an Upper Sedimentary Group above the lavas. A revision in progress will replace these terms with formal lithostratigraphical divisions.

The Cementstone Group consists of interbedded cementstones, mudstones and sandstones (Plate 9.1). The cementstones are pale grey, fine-grained,

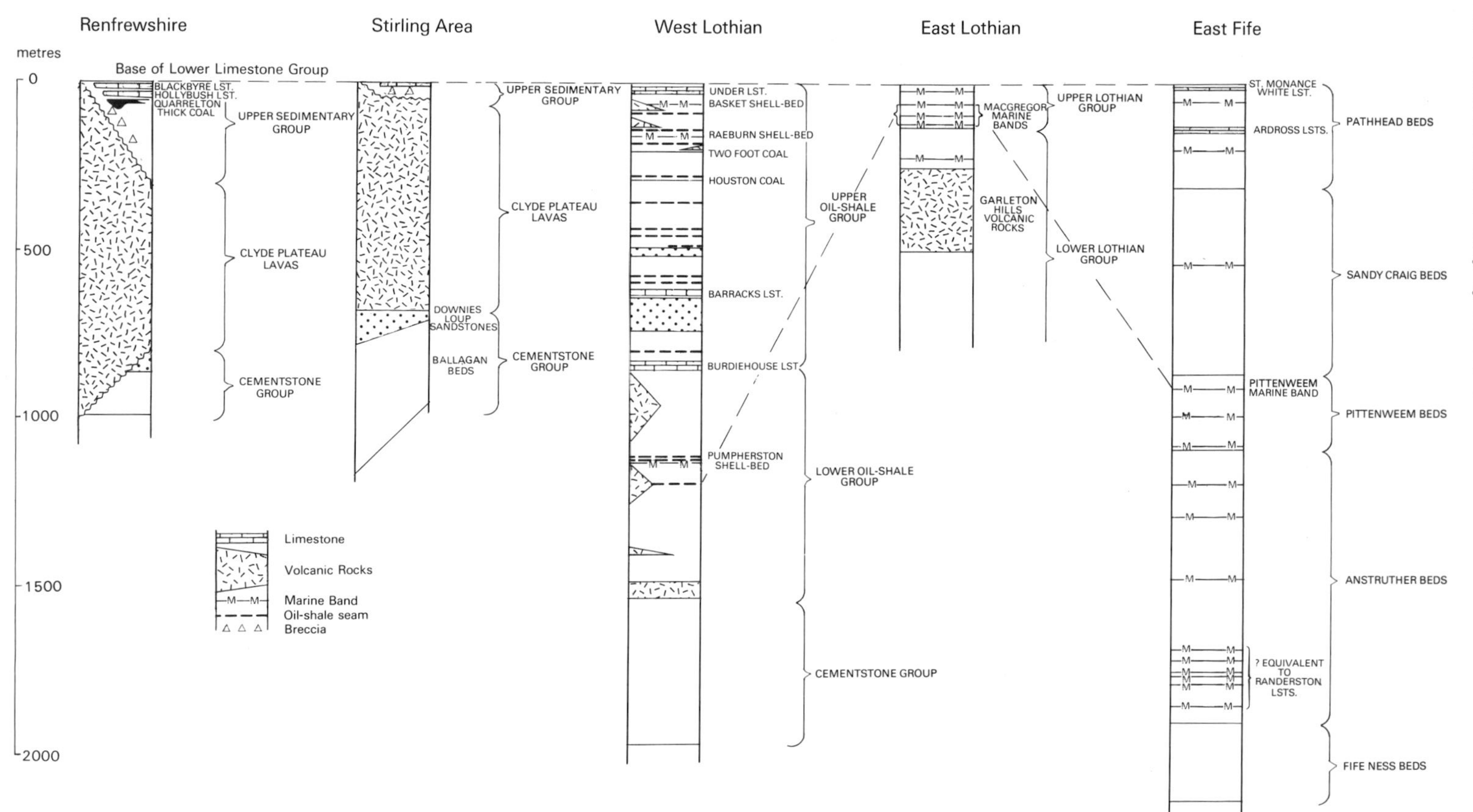

Figure 18 Comparative generalised vertical sections of the Calciferous Sandstone Measures

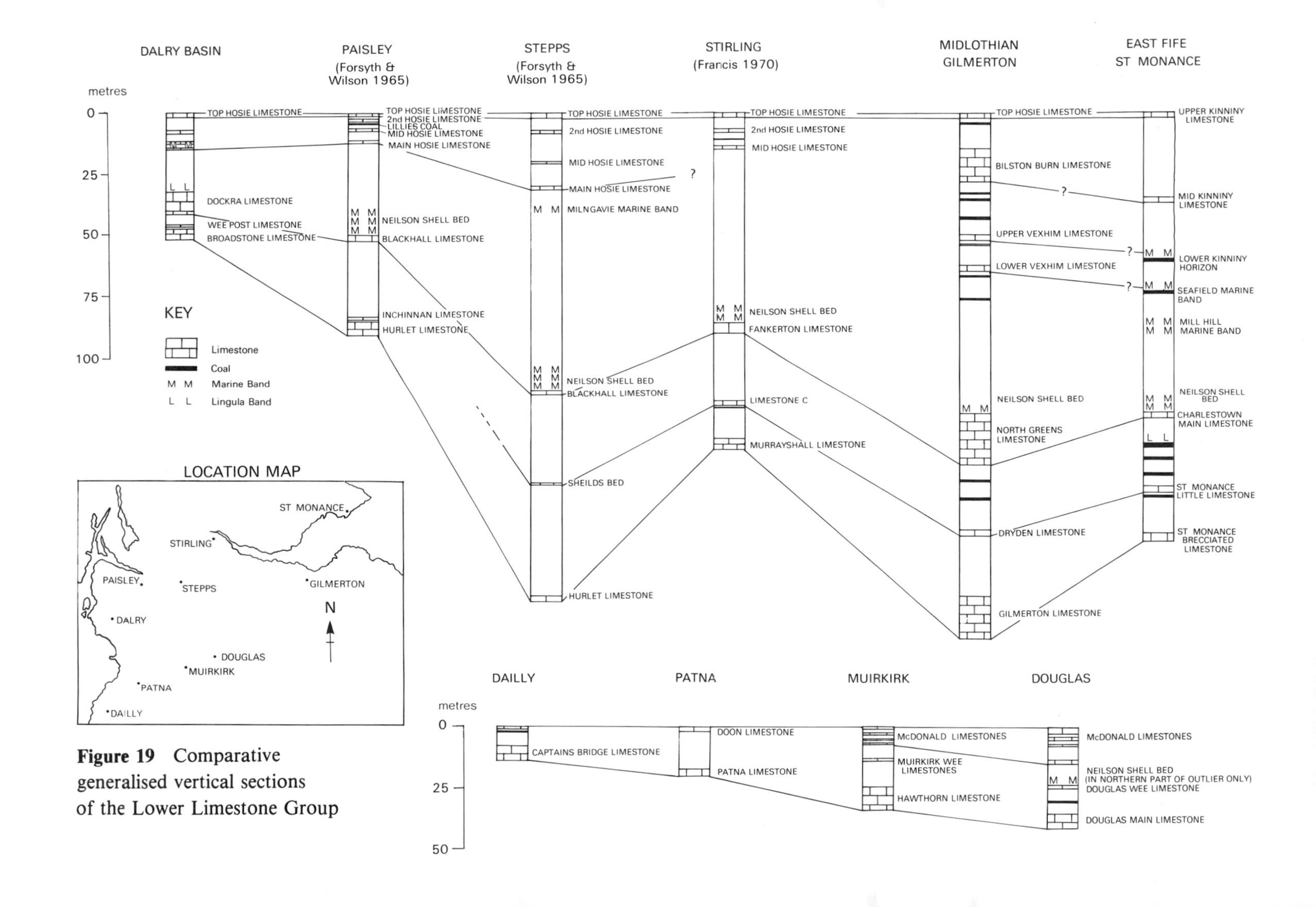

Figure 19 Comparative generalised vertical sections of the Lower Limestone Group

dolomitic limestones containing a variable amount of mud, silt and fine sand. Individual beds occur up to 1 m thick, but are usually less than 0.3 m thick and are of uniform thickness. They occur in groups interbedded with mudstones.

Cementstones are broadly of two types. Layered cementstones have a sharply defined top and base and some may show signs of internal stratification. They may also have sedimentary features and desiccation cracks. Nodular cementstones show a transition into calcareous mudstones above and below and may form a more or less persistent layer or a row of nodules. The layered cementstones are thought to be primary in origin and the nodular cementstones of secondary origin.

The mudstones interbedded with the cementstones are dark grey or greenish grey, and are mottled brown or purplish brown in places. They are calcareous, silty, poorly bedded and range in thickness from a few millimetres to several metres.

Sandstones and siltstones also occur in the sequence and are very variable.

The Cementstone Group can be subdivided in some areas into a lower division consisting of a cementstone sequence and an upper division which is a fluviatile sandstone. The cementstones around the Campsie Fells are named the Ballagan Beds. The overlying fluviatile sandstone resembles in some respects the Cornstone Beds of the Upper Old Red Sandstone and is known as the Spout of Ballagan Sandstone. It is probably equivalent to the Downie's Loup Sandstones in outcrops west of Stirling.

The thickness of the Cementstone Group varies considerably in both its divisions. The Ballagan Beds range in thickness up to about 380 m and the sandstone varies from about 30 to 100 m. Locally one or both divisions are absent and the lavas rest directly on Upper Old Red Sandstone facies rocks.

The Cementstone Group is succeeded by the Clyde Plateau lavas whose outcrop forms a semi-circle of hills extending from Stirling south-westwards to Greenock and from there south-eastwards to Strathaven. The lavas are also known to underlie the Central Coalfield at least as far east as Salsburgh. The lava pile is up to 900 m thick. It rests with apparent conformity on the Downie's Loup Sandstones which contain volcanic detritus in the upper part.

The lavas are mainly olivine-basalts with some trachytes and they are described in greater detail in Chapter 12.

The Upper Sedimentary Group rests on an irregular, weathered surface of the lavas. In many places the earliest sediments are conglomerates, sandstones and mudstones composed mainly of volcanic material. The upper part of the group consists of mudstones, sandstones, coals and up to three marine limestones.

The variation in thickness of the group reflects the topography of the lava surface on which it was deposited. Some areas of the lava outcrop remained above the level of deposition until Lower Limestone Group times, but in other areas a considerable thickness of sediment accumulated.

The thickest sequence occurs in the Paisley area where it is about 360 m thick. The lower part of the succession consists of volcanic detritus which is overlain by sandstones with some coals. The upper part of the sequence consists of mudstones, sandstones, seatclays and coals with up to three thin limestones. Two of the limestones, the Blackbyre and the Hollybush, attain a reasonable thickness and persistence and have been correlated over much of the region.

An extraordinarily thick coal occurs in the sequence to the south-west of Paisley. Several seams thicken and come together by attenuation of the intervening strata to form the Quarrelton Thick Coal. It is up to 15 m thick but is locally doubled in thickness, perhaps by contemporaneous gravity sliding. The coal is riddled with old workings some of which were as early as 1634.

North of the Clyde, in the Campsie area, the general succession is similar but rather thinner. It is exceptional in that the lower part of the sequence consists of a quartz-conglomerate and sandstone which is largely free of volcanic detritus.

In north Ayrshire, the sequence above the lavas probably does not exceed about 30 m, but two limestones are present and are correlated with those in the Paisley district.

South-west Midland Valley

In the southern part of the Midland Valley, between south Ayrshire and Douglas, the Calciferous Sandstone Measures consists of red, white, yellow or pink sandstones, red and grey-green mudstones with subordinate conglomeratic beds and rarely thin impersistent coals with seat clays. In places cornstone nodules occur and the predominantly arenaceous sequence is locally interdigitated with the cementstone-shale facies similar to the Ballagan Beds. The strata are not well exposed. They are apparently marginal to the main area of sedimentation and were possibly also affected by contemporaneous fault movement.

In the Douglas area the upper part of the sequence contains two marine limestones which can be correlated with the Blackbyre and Hollybush limestones of the Paisley area. At Muirkirk there is only one limestone in the corresponding part of the succession. More detailed correlation between sequences below the Lower Limestone Group is not possible.

One of the thickest sequences in this area occurs around Dailly where the thickness is estimated to be about 510 m. However, the thickness tapers off to a few metres in parts of the Douglas area south of the Kennox Fault where only the uppermost beds were deposited.

Edinburgh area and West Lothian

The Calciferous Sandstone Measures in this area are made up of the Cementstone Group and the Lower and Upper Oil-Shale groups and the sequence is one of the thickest in the region.

The base of the Cementstone Group is taken at the lowest occurrence of the cementstone-shale facies. The sediments consist of sandstones and sandy shales, cementstones and mudstones and locally conglomerates at the base. The sandstones may be red, brown, grey or white and the argillaceous beds red, brown or greenish grey. Nodular cornstones occur in places and there are thin layers of gypsum in some beds. The sequence is predominantly arenaceous but there is an interdigitation of the sandstone and cornstone facies with the cementstone-shale facies. The thickness ranges from about 760 to about 1120 m.

The Lower Oil-Shale Group consists of the Arthur's Seat Volcanic Rocks at the base overlain by a sequence consisting principally of pale sandstones and dark grey shales. The thickness of the group has been calculated to be over 1000 m in Midlothian but thins to about 700 m in West Lothian.

The sediments overlying the Arthur's Seat Volcanic Rocks consist of dark grey shales, bituminous in places, with ironstone nodules and interbedded with mainly fine-grained, pale yellow, grey or pinkish sandstones. The thicker sandstone units are named the Craigleith, Ravelston and Hailes Sandstones and were formerly worked for building stone. Thin limestones and rare, thin, impersistent coal seams also occur.

In the upper part of the sequence oil-shales are developed at two horizons. The more important development is known as the Pumpherston Oil-Shales which were formerly mined in West Lothian. Immediately underlying them is the Pumpherston Shell Bed which is the only horizon in the Lower Oil-Shale Group in this area which contains a fauna which enables correlation with other successions.

The Upper Oil-Shale Group extends from the base of the Burdiehouse Limestone to the base of the Lower Limestone Group and is characterised by the occurrence of nine or ten oil-shale seams which were formerly mined extensively in West Lothian. The group is thickest around West Calder in Midlothian where it is about 850 m thick but it thins eastwards and southwards with a reduction in the number of oil-shale seams. In the west side of the Midlothian coalfield it is about 450 m thick and to the south around Carlops it is about 150 m thick.

The Burdiehouse Limestone at the base of the Group is up to 15 m thick and is thought to be of freshwater or estuarine origin. Its fossil content of plants, fish and ostracods is not diagnostic and its lateral equivalence to other limestones is based mainly on lithological similarity.

Most of the sequence consists of argillaceous rocks and contains seams of oil-shale up to 5 m thick. They are hard and usually very dark brown in colour and are minutely laminated. They tend to grade into bituminous or carbonaceous shale.

The sequence also includes the Houston Coal and the Two Foot Coal. They are both thin and of inferior quality and only the former was mined.

Thick beds of 'marls' occur at several levels in the sequence. They are very poorly bedded mudstones with thin calcareous bands and are greenish grey in colour. Thick sandstones are also present in the lower part of the sequence and were worked for building stone at one time.

In the upper part of the Upper Oil-Shale Group there are three horizons containing marine fossils which are useful for correlation. The oldest of the three is the Raeburn Shell Bed which occurs below the Raeburn Shale and is widely distributed in West Lothian and parts of Midlothian. Higher in the sequence a marine band in West Lothian is correlated with the Basket or Cot Castle Shell Bed of Lanarkshire and with the Cephalopod Limestone of the Midlothian Coalfield. The third marine horizon is thought to be equivalent to the Under Limestone of Lanarkshire and the Bone Bed Limestone in Midlothian.

Beds of volcanic ash and lava occur at several horizons in the Lower and Upper Oil-Shale groups in addition to the Arthur's Seat Volcanic Rocks at the base. The lavas of the Bathgate Hills make their earliest appearance just above the Two Foot Coal in the Upper Oil-Shale Group and they extend up into the Upper Limestone Group of that area.

East Lothian

In East Lothian the Calciferous Sandstone Measures are subdivided into the Lower and Upper Lothian groups. The division is made at the base of a group of marine bands known as the Macgregor Marine Bands which are thought to be equivalent to the Pumpherston Shell Bed in West Lothian.

The Lower Lothian Group is predominantly arenaceous, red at the base but interdigitated with cementstone-shale facies at at least two levels. It also includes the Garleton Hills Volcanic Rocks. The Upper Lothian Group includes several marine limestones and the Macgregor Marine Bands. It also includes a few thin coals but oil-shale seams are absent.

The Upper Lothian Group is about 140 m thick and the Lower Lothian Group is at least 730 m thick near Pencaitland.

East Fife

The Calciferous Sandstone Measures in east Fife have been subdivided into an ascending succession of Fife Ness, Anstruther, Pittenweem, Sandy Craig and Pathhead beds. The strata consist mainly of pale sandstones, grey mudstones and siltstones, many showing faunal evidence of non-marine deposition.

Numerous thin impersistent coal seams and seat beds are also developed and thin dolomitic limestones occur throughout the sequence.

Marine limestones and mudstones are present throughout the succession and become more common in the upper part of the sequence.

The maximum known thickness is over 2000 m, but miospore evidence indicates that the oldest beds seen are of Viséan age. It is possible that a basal cementstone facies with the older Tournaisian miospore zones is present at depth.

Lower Limestone Group

The Lower Limestone Group is the uppermost subdivision of the Dinantian in Scotland. It follows the Calciferous Sandstone Measures conformably, except in some marginal areas where it overlaps them and rests on older strata.

The marine influence, seen in its early phases in the upper part of the Calciferous Sandstone Measures, is more fully expressed in the Lower Limestone Group. The proportion of limestone in the sequence is greater in the Lower Limestone Group than it is in the other subdivisions of the Carboniferous, although it is still a relatively minor constituent.

The base of the Group is placed at the base of the Hurlet Limestone or its lateral equivalents. It is a widespread horizon and its correlation has been agreed over most of the region. The top of the Group is taken at the top of the Top Hosie Limestone. The latter is present in most outcrops in the Midland Valley and it can be recognised with little difficulty (Figure 19).

The group has its greatest development in the Midlothian–east Fife area where it is up to 220 m thick.

Lithology

The rock-types of the Lower Limestone Group consist principally of sandstone, mudstone, limestone and coals with root-beds. The sequence is cyclic and the marine members tend to be the best developed, most widespread and persistent horizons.

The sandstones, which are the most prevalent rock-type in the thicker parts of the outcrop, are mainly fine- or medium-grained and grey or pale yellow in colour. However, coarser, locally pebbly, sandstones occur which may represent channel-fill deposits.

The argillaceous strata are dark grey, silty and contain thin ironstone bands which were worked to a small extent. Mudstones occurring above and below the limestone horizons tend to be marine and at several horizons contain varied marine faunas dominated by brachiopods and molluscs.

The limestones occur in beds up to 20 m thick but are commonly much thinner. They tend to be hard, argillaceous, grey limestones, which, at some horizons, have a conspicuous fauna of corals, brachiopods and crinoids. Some are partially dolomitised locally. The limestones normally contain a marine fauna, but the lower part of the Blackhall Limestone in the Glasgow area contains a fauna of ostracods and fish-debris and has been described as non-marine.

Coal seams were developed irregularly, but a few were thick enough to have been worked to some extent in parts of the Central Coalfield and in Midlothian and Fife. The Lillies Shale Coal in the Glasgow area was worked at one time as an oil-shale.

Lateral variation

Major variations in thickness in the Lower Limestone Group can be attributed to differential subsidence in basinal areas whose disposition is controlled to some extent by the presence of thick lava piles in the Calciferous Sandstone Measures. In addition there was differential subsidence across lines of faulting particularly in Ayrshire, and overlap and attenuation occurred at the limits of the depositional area.

The greatest accumulation of sediment occurred in the Central Coalfield–Kincardine basin area and in the Midlothian–east Fife basin, where thicknesses up to 220 m are known. These two areas are separated by a NNE-trending zone of relative thinning called the Burntisland Anticline, which coincides with and is probably a consequence of a thick pile of lavas which extends from the Calciferous Sandstone Measures up into the Upper Limestone Group. The Lower Limestone Group sediments in the Central Coalfield–Kincardine basin area show marked attenuation towards the arcuate outcrop of the Clyde Plateau lavas extending from the Kilsyth Hills to Greenock and south-eastwards to Strathaven. At least part of the high ground between the Central and Ayrshire coalfields was an area of non-deposition.

In Ayrshire the strata are seldom over 60 m thick and variations in thickness are abrupt and coincident with north-easterly trending lines of faulting. In the Dalry Basin in the north the sediments are up to about 60 m thick, but are locally thinned by overlap against the eroded surfaces of the Clyde Plateau lavas. The thickness is reduced to between 20 and 30 m on the south side of the Dusk Water Fault. Between the Dusk Water Fault and the Inchgotrick Fault the strata increase in thickness southwards to about 50 m, but thin towards the north-east and are cut out by overlap against the lavas north of Galston.

In the area on the south side of the Inchgotrick Fault the strata are thin in the west and absent further east. The strata, where present, are not only thin but parts of the succession are missing either through non-deposition or contemporaneous erosion. The area to the south of Ayr is thought to have been emergent during Lower Limestone Group times.

The Kerse Loch Fault is also a line of faulting across which there is an abrupt increase in thickness. The thickness north of the fault varies between 6 and 12 m and south of the fault it is between 16 and 22 m.

In the Central Coalfield similar differential subsidence is thought to have occurred on the south side of the Campsie Fault.

The lithological variation in the Lower Limestone Group is considerable and is to some extent linked to the pattern of subsiding basins. The proportion of sandstone in the sequence is greater in areas of greater thickness, and in thinner sequences the limestones and shales are more prominent.

Lateral variation also occurs in the composition of the marine faunas which may be dependent on their lithological association, which in turn relates to the pattern of differential subsidence. Certain species occur only in areas of reduced sedimentation which were receiving deposits mainly of calcareous shales and limestones and these species apparently did not occur in the main areas of subsidence where there was a great proportion of clastic material laid down.

Contemporaneous volcanic rocks

Volcanic rocks occur in the Lower Limestone Group succession in parts of West Lothian and Fife. Between Bo'ness and Bathgate the sequence is made up almost entirely of basalt lavas and tuffs with some sedimentary intercalation including limestones. Part of the sequence in the Kinghorn area of Fife is also taken up by lavas and tuffs.

8. Namurian

The Namurian strata present in the Midland Valley are divided on lithological criteria into the Limestone Coal Group, the Upper Limestone Group and the Passage Group, which together make up the Millstone Grit Series (Figure 20).

The strata consist of shallow water, mainly terrigenous sediments deposited in cycles in a subsiding fluvio-deltaic environment. The Limestone Coal Group is characterised by the presence of numerous coals and the relative scarcity of marine strata compared with the underlying Lower Limestone Group. The Upper Limestone Group contains major marine cycles with thick limestones developed in some areas but much of the group is composed of non-marine strata including some coals. The Passage Group, formerly the Scottish Millstone Grit, consists mainly of sandstones and clay-rocks, some of the

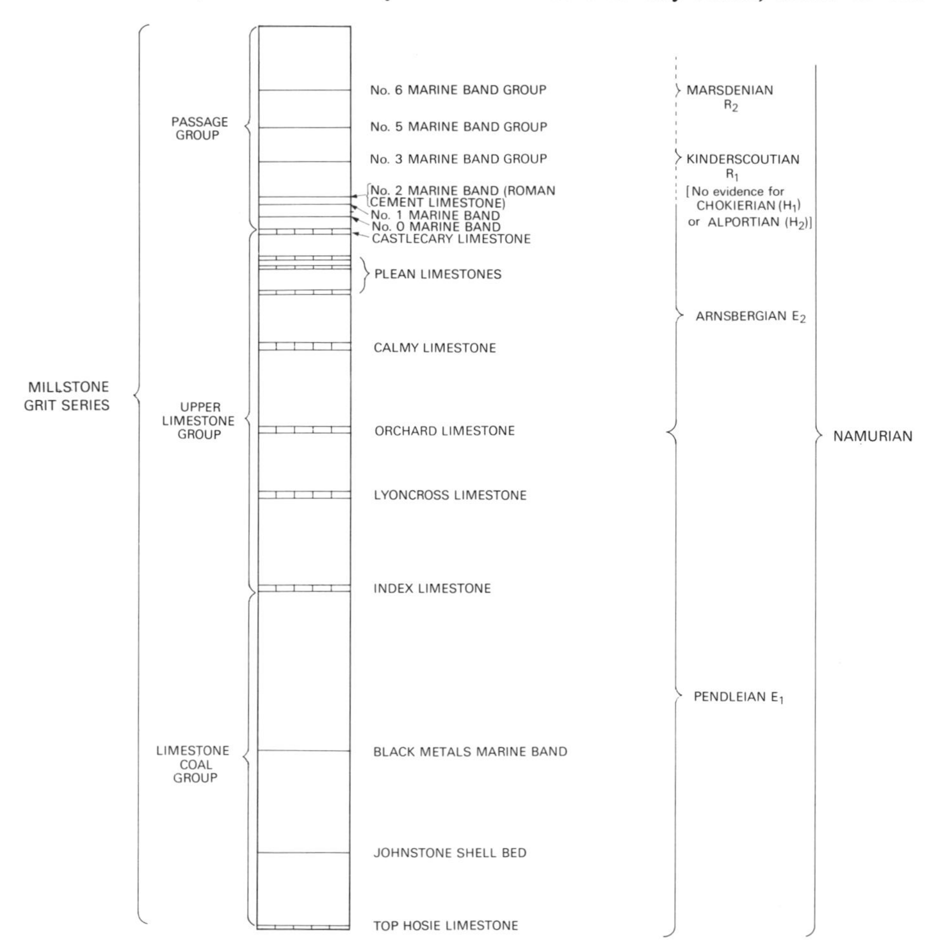

Figure 20 Generalised vertical section of the Namurian in the Midland Valley

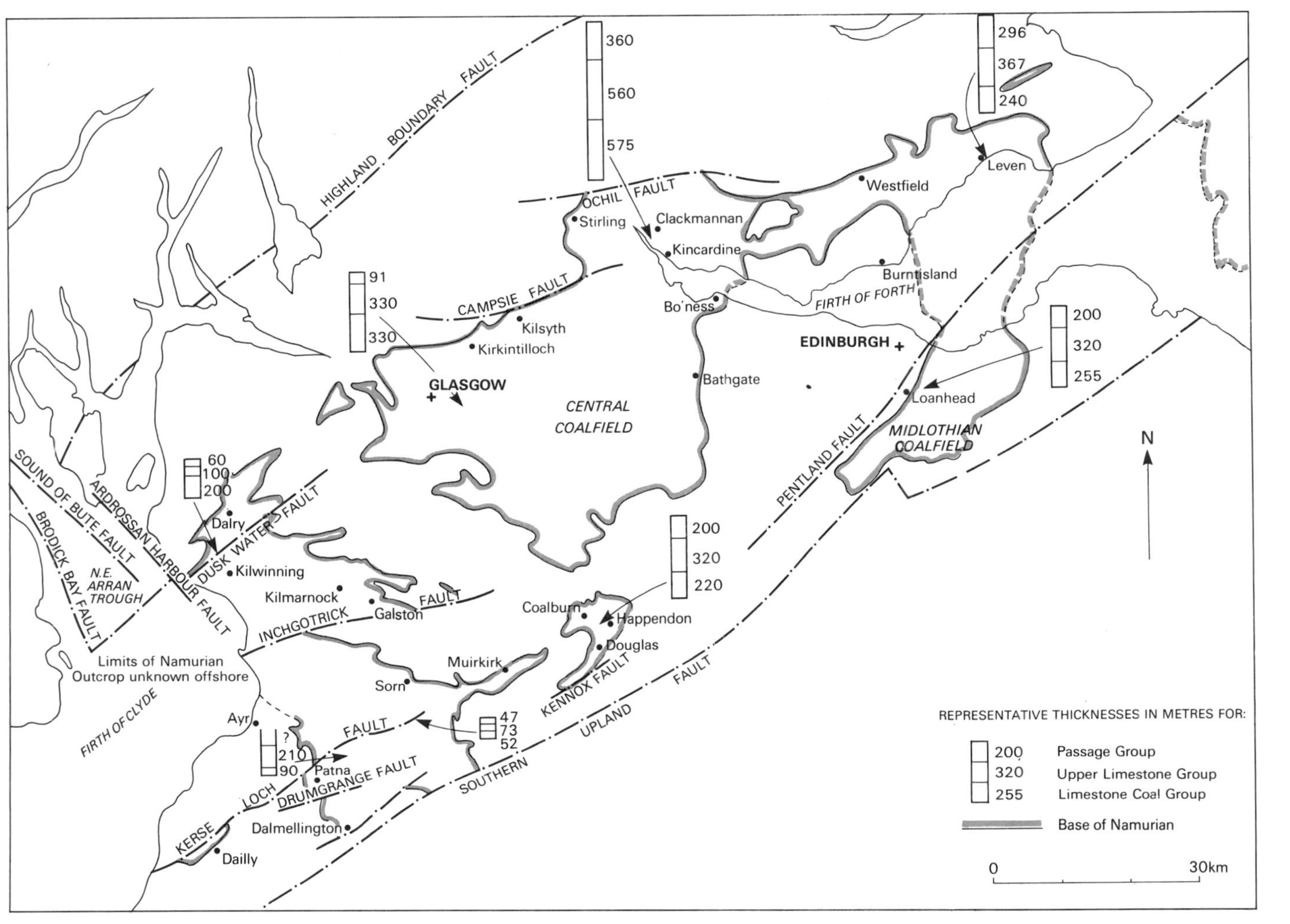

Figure 21 Distribution of Namurian strata in the Midland Valley. In the Firth of Forth the position of the base of the Namurian is drawn from geophysical data and some boreholes and is only approximate. The limits of the Namurian in the Firth of Clyde are unknown

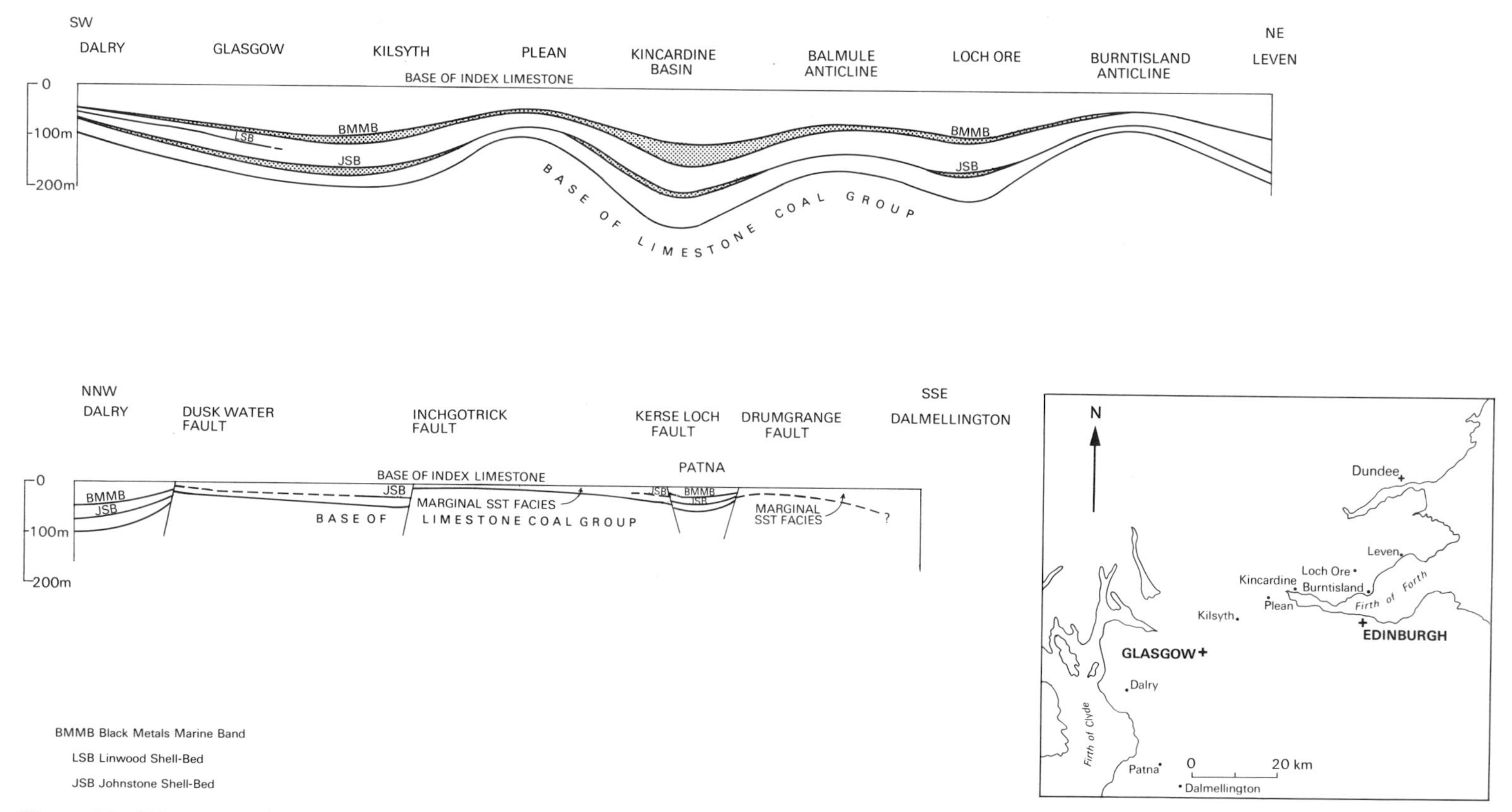

Figure 22 Diagrammatic cross-sections to illustrate variation in thickness in the Limestone Coal Group

latter containing rootlets. Marine bands occur in the lower part of the sequence with the richer faunas in the lowest bands. Coals are present but are mostly thin and impersistent. Unconformities are evident at two levels at least, indicating differential uplift and erosion.

In England the Namurian is classified into stages based on the vertical distribution of goniatite species. The lowest two of these stages, the Pendleian (E_1) and the Arnsbergian (E_2) are well developed in Central Scotland but the later ones are poorly represented or have not been recognised. The distribution of Namurian strata is shown on Figure 21.

Limestone Coal Group

The Limestone Coal Group varies greatly in its development from a thin sandy sequence about 30 m thick with thin coals in parts of Ayrshire, up to 550 m in the northern part of the Central Coalfield, where there are about fifteen workable coal seams (Figure 22). A large part of the Scottish coal-mining industry has been, and continues to be, based on the coals of the Limestone Coal Group.

The base of the group is taken at the top of the Top Hosie Limestone, the uppermost bed of the Lower Limestone Group. The top of the group is drawn at the base of the Index Limestone so named because it indicated that the valuable coals lay underneath. The strata are all of Pendleian (E_1) age.

Lithology

The strata consist principally of sandstones, siltstones and mudstones with seatearths or seatclays and coals. The sandstones are normally fine-grained and grey or pale yellow in colour but coarse-grained pebbly sandstones occur in parts of the sequence. The latter are channel-fill deposits with an erosive transgressive base which may cut down into and replace older deposits. The mudstones are grey to dark grey, locally carbonaceous, often silty and in places inter-laminated with siltstone or fine sandstone. Seatearths and seatclays are fossil soils penetrated by the roots of the vegetation which gave rise to the coal seams. The sediment consists of sandstone or mudstone and the bedding may remain obvious or be destroyed by soil-forming processes. Small ironstone nodules are common.

Nodular and bedded clayband ironstones occur within the more argillaceous parts of the succession and were formerly worked. Blackband ironstones were also of economic importance in the early years of the iron and steel industry in Scotland. These consist of iron carbonate with enough coaly material to make them virtually self-smelting.

Coal seams are rather variable in their development. At their thickest they exceed 2 m but this occurs only locally. There are, however, many seams over 1 m thick. Correlation of coals from one coalfield to another is seldom possible as the rapid changes in thickness, and splitting of seams makes comparison difficult. This difficulty has resulted in the use of different sets of names for coals in the same part of the succession in adjacent areas.

Although the sediments are mainly of fluvio-deltaic, non-marine origin, two marine incursions affected most of the region. These are recorded by the Johnstone Shell Bed and the Black Metals Marine Band which occur in the lower part and about the middle of the group respectively. Relatively rich marine assemblages are present in both marine bands over most of the region but in some marginal areas the fauna is reduced to *Lingula* or is absent (Figure 22). An extra marine band also occurs between the major ones in north Ayrshire and the Glasgow area and numerous bands with *Lingula* only are present in most areas. The marine fossils are mainly found in mudstones and more rarely in ironstones.

Correlation within the group is based on the two principal marine bands and the *Lingula* bands are useful for local correlations.

Lateral variation

Lateral variation within the group is considerable both in thickness of sediment between one area and another and in the lithological sequence (Figures 21, 22). Differences in thickness are due to differential subsidence in sedimentary basins and contemporaneous movement on north-easterly trending faults. The differences in the sediments record the changing environment of deposition within the delta area.

In the Glasgow area and north Ayrshire the lower part of the group consists of a relatively high proportion of mudstone compared to sandstone, and clayband ironstone occurs at several levels. The coals are relatively few and thin, but the two main marine bands are present and a third marine horizon is also present in at least part of the area.

The upper part of the group contains a greater proportion of sandstone, and coals are more numerous and thicker. There are also about 15 bands containing *Lingula* above the Black Metals Marine Band.

As the strata are traced eastwards towards the Kilsyth area, the group as a whole increases in thickness from about 200 m in north Ayrshire to 330 m in the Glasgow area and to 400 m near Kilsyth. The increase in thickness is mainly due to additional beds of siltstone and sandstone in the succession. The number of coals also increases.

In the west, the Johnstone Shell Bed consists of about 22 m of dark mudstones with a varied marine fauna. Further east the shell bed splits into separate leaves of fossiliferous mudstone separated by beds of siltstone and sandstone. Rooty horizons and thin coals have formed locally at the top of the sandy beds. In the Kilsyth area, the Johnstone Shell Bed is split into as many as five leaves. The richest fauna is restricted to the bottom leaf and the fauna in the other leaves is reduced to *Lingula* and bivalves.

Lateral changes in thickness are also seen in the Kincardine Basin where the Limestone Coal Group reaches its maximum thickness of about 550 m. The proportion of sandstone, particularly in the lower part of the sequence, increases across the region towards the east as does the total thickness of coal. There is also a decrease in the number of *Lingula* bands in the sequence from west to east. In the Kincardine Basin *Lingula* bands are more common in the central parts of the basin than they are on the flanks.

In Fife and Midlothian the group contains a considerable proportion of sandstone and there are numerous coals, many of economic thickness throughout the succession. The Johnstone Shell Bed and the Black Metals Marine Band are both present but in places they are split into two leaves by a bed of sandstone. The number of *Lingula* bands is reduced to about eight in Midlothian and four or five in the Leven area of Fife.

In Ayrshire, south of the Dusk Water Fault, the Limestone Coal Group succession is attenuated and in places shows a marginal development. Sedimentation has been controlled by contemporaneous movement on north-easterly trending faults. The most important of these are the Dusk Water, Inchgotrick, Kerse Loch, Drumgrange and Southern Upland faults.

The thickness of the group is reduced to about 50 m on the south side of the Dusk Water Fault near Kilwinning, although this increases to about 90 m north of Kilmarnock. Coals, thick enough to have been worked, are only locally developed. North of Darvel, the base of the group overlaps the Lower Limestone Group and rests unconformably on Lower Carboniferous lavas. At several localities in the outcrop from Kilwinning eastwards, a coarse erosive sandstone in the overlying Upper Limestone Group has eroded the Index Limestone and rests unconformably on rocks in the upper part of the Limestone Coal Group.

Further attenuation occurs on the south side of the Inchgotrick Fault. The group is represented in this area by as little as 15 m of sandstone which is pebbly in places and locally reddened. The reddening suggests local contemporaneous emergence. The sequence thickens towards the south-east where the pattern of cyclic sedimentation becomes re-established and the two marine bands and several workable coals appear in the succession at Sorn. In the vicinity of Ayr it is probable that there was no deposition during Limestone Coal Group times.

In the area south-east of Galston, the Limestone Coal Group strata rest on rocks of Upper Old Red Sandstone facies.

There is a marked increase in thickness in the strata on the south side of the Kerse Loch Fault. The strata change from marginal sandstone facies on the north side of the fault to coal-bearing deltaic cycles on the south side. In the Dailly coalfield, the sediments are predominantly arenaceous with several workable coals. Further north around Patna, where the group is about 80 m thick, the sequence consists of cyclic repetitions of sandstones, mudstones and coals. The two marine bands contain only *Lingula*, but locally the Black Metals Marine Band contains a more varied fauna. In Dailly a *Lingula* band is perhaps the local equivalent of the Black Metals Marine Band.

The strata thin from Patna towards the south and south-east and the facies reverts from coal-bearing cycles to marginal sandstones and conglomerates which are reddened in places. The thinning is abrupt across the line of the Drumgrange Fault south-east of Patna.

The arenaceous facies around Dalmellington indicates proximity to the margin of the depositional area during Limestone Coal Group times. South-west of Dalmellington the base of the sequence rests on Lower Devonian lavas.

At Muirkirk the maximum thickness of the group is about 110 m, but this is reduced to 34 m a few miles to the south-west. Several coals of economic importance are developed in the sequence, but they are fewer and thinner where the group as a whole is thinner. The Johnstone Shell Bed and the Black Metals Marine Band are both present throughout the area, but the latter contains little more than *Lingula* in areas where the succession is attenuated.

In the Douglas area the Limestone Coal Group ranges in thickness from 60 m in the south-west part of the coalfield to 220 m in the north-east. The development in the north-east contains eight workable coals, but this is reduced to five on the west side of a north-easterly trending zone of minimum thickness. To the south-west the coal content is further reduced in thickness and the sequence is more arenaceous. On the south-east side of the Kennox Fault erosion in later Carboniferous times caused younger Carboniferous sediments to overstep a thin development of Limestone Coal Group strata.

The Johnstone Shell Bed is present throughout the Douglas coalfield but the Black Metals Marine Band is poorly developed and contains only *Lingula*.

Contemporaneous volcanic rocks

In north Ayrshire, in an area west and south of Dalry, south of the Dusk Water Fault, beds of volcanic tuffs, up to 25 m thick are interbedded with normal sediments at several levels. In places the Dalry Blackband Ironstone is replaced laterally by volcanic tuffs.

In Fife similar pyroclastic rocks and, more rarely, lavas occur locally in the succession.

A sequence of basalt lavas in the Bathgate to Linlithgow area of West Lothian replaces most of the Limestone Coal Group sediments and in the Bo'ness area lavas interdigitate with the sedimentary rocks.

Upper Limestone Group

The Upper Limestone Group strata show a return to similar marine conditions as existed during Lower Limestone Group times and which were absent during the deposition of the intervening Limestone Coal Group. This is marked by the return of thick limestones and beds of mudstone containing rich and varied marine faunas. Coals are generally poorly developed, in comparison with those of the Limestone Coal Group, but thick beds of sandstone are a feature of the group, particularly in the lower part. The group attains a maximum development of about 590 m in the Kincardine Basin.

The limits of the group are from the base of the Index Limestone up to the top of the Castlecary Limestone. In relation to the goniatite zonation the group is partly in the Pendleian (E_1) Stage and partly in the Arnsbergian (E_2) Stage.

Lithology

The Upper Limestone Group succession is cyclothemic in character. There are up to nine marine limestones in the group with varying degrees of development and persistence throughout the area (Figure 23). The thickest and most persistent of these beds are the Index, Orchard, Calmy and Castlecary limestones, in ascending order. The other limestones, which are not everywhere developed are the Huntershill Cement, Lyoncross and Pleans Nos. 1, 2 and 3 limestones.

The Index Limestone, at the base of the group, is a fine-grained grey or brownish grey rock usually rather argillaceous at the top and bottom. Its thickness varies between 1 and 2 m but it is locally thicker in the Muirkirk and Douglas areas. The limestone has not been found in east Fife and in some parts of north Ayrshire it has been removed by pene-contemporaneous erosion.

The Orchard Limestone shows more lateral variation than the Index Limestone. In north Ayrshire and the Glasgow area it consists of well-bedded, fine-grained, dark argillaceous limestone, split in places into two or three beds. In Midlothian and Fife, the same horizon is called the Orchard Beds and they

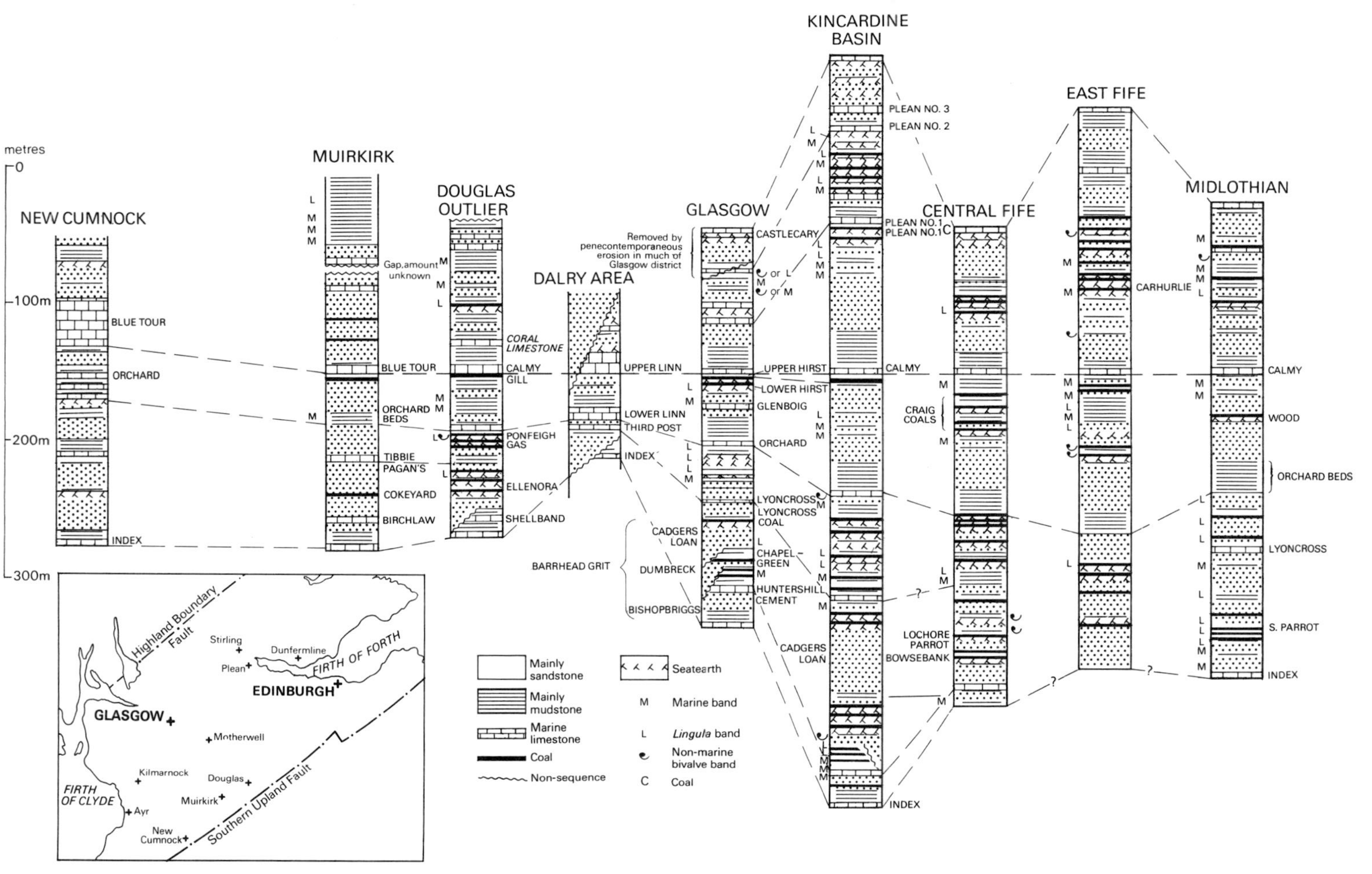

Figure 23 Comparative generalised vertical sections of the Upper Limestone Group

consist of fossiliferous marine shales which locally include thin limestone bands. The Orchard Limestone in north Ayrshire, where it is called the Lower Linn Limestone, is about 6 m thick and the Orchard Beds of Midlothian are about 15 m thick. The horizon can be recognised in successions throughout the outcrop in the Midland Valley.

The Calmy Limestone lies about the middle of the Upper Limestone Group in complete sequences. Typically it is pale grey and argillaceous (hence the name), usually occurring in two beds, but locally three or four. The limestone itself contains only fragmented brachiopod shells and crinoid columnals but the shales above and below are more fossiliferous. The limestone is well developed in north Ayrshire where it reaches about 15 m and is known as the Upper Linn Limestone. An exceptionally thick development of about 30 m occurs locally in south Ayrshire. There are also thick developments in the Muirkirk and Douglas areas. Elsewhere, it is usually less than 4 m thick.

Near the base of the mudstones underlying the Calmy Limestone, there is a thin bed of carbonaceous shale in which the bivalve *Edmondia punctatella* occurs in abundance. The *Edmondia punctatella* Band is only a few centimetres thick in central and western areas but it has a very wide distribution in the Midland Valley.

The Castlecary Limestone marks the top of the Upper Limestone Group, but in many places in the Midland Valley it has been removed by erosion prior to deposition of the Passage Group. It is not present in Ayrshire, nor in the Muirkirk and Douglas areas. It is also absent in parts of the Central, Fife and Midlothian coalfield areas. The limestone ranges in thickness from 1.5 to 5.5 m and tends to occur in at least two beds. The character of the limestone varies from fine-grained to crystalline and is commonly dolomitised. The rock is mottled light and dark grey in places, but north-east of Glasgow, contemporaneous weathering has reduced the limestone to a green clay and calcareous nodules.

All the limestones, except the Castlecary, are overlain by shales with a marine fauna. Despite the obvious geological interest and correlative importance of the limestones, they form only a very small proportion of the lithological sequence. The predominant rock type of the group is sandstone and considerable thicknesses are developed at several levels. These sandstones, which are usually white to pale yellow or pale grey in colour, frequently have an erosive base and rest unconformably on the underlying strata. In the Glasgow area, one such sandstone, the Barrhead Grit, originates in the sequence above the Huntershill Cement Limestone, but it transgresses downwards in the sequence and rests unconformably on strata below the limestone. In north Ayrshire a coarse sandstone at a similar horizon locally cuts out the Index Limestone.

The coals of the Upper Limestone Group are mostly thin but some have been worked locally. The Upper Hirst Coal, which occurs below the Calmy Limestone, is worked extensively in the Kincardine Basin where it is up to 2.5 m thick. It has a rather high ash content which makes it suitable for power station use.

Correlation

The four main limestones which can be recognised throughout most of the Midland Valley form the essential framework of correlation and they are

1 Ballagan Beds, overlain by sandstones equivalent to Spout of Ballagan Sandstone, Stirling, Stirlingshire. Interstratified thin cementstones and mudstones. (D 1881)

2 Western Campsie Fells from Strathblane. Clyde Plateau Lavas form stepped scarp above Lower Carboniferous sediments partly covered by landslip. On left, prominent hills, Dumgoyne and Dumfoyne, are volcanic vents, near-contemporaneous with lavas. (D 2995)

Plate 9

1 St Monance Syncline in the Lower Limestone Group sediments of east Fife. (D 2796)

Plate 10

2 Base of late-Carboniferous quartz-dolerite sill intruded into sandstones of the Calciferous Sandstone Measures, Hound Point, South Queensferry. (D 1917)

supplemented by up to five other less persistent limestones. In addition, there are numerous other marine horizons with a more restricted fauna. These tend to be better developed where the sequence is thickest and are poorly developed or absent in areas where the group is attenuated.

Individual limestones are recognised by a combination of lithological and faunal characteristics. In many cases the faunal content of the limestones themselves is rather poor, the richest and most abundant faunas occurring in the mudstones associated with limestones.

The fossil assemblages show lateral changes which are more or less consistent at each horizon and these may be related to differing rates of subsidence and sedimentation. The differences are evident between east and west on either side of the Burntisland Anticline and there are also variations between south-east and north-west affecting the lower part of the group in the Central Coalfield.

Lateral variation

The pattern of variation in thickness in the Upper Limestone Group is similar to that of the Limestone Coal Group. The maximum thickness occurs in the Kincardine Basin and there is notable thinning over the Burntisland Anticline. In Ayrshire, north-easterly trending faults exerted a control on the thickness of preserved sediment in the same manner as occurred in the Limestone Coal Group, but to a lesser extent.

The greatest known thickness of sediment in the group occurs near Clackmannan, in the Kincardine Basin where the complete succession is about 590 m thick. All the previously mentioned limestone horizons are developed including the three Plean Limestones. In addition there are numerous other beds of marine shale and *Lingula* bands.

In the lower part of the succession there are coarse erosive sandstones which locally cut down through older beds. Similar sandstones occur at comparable levels in the Glasgow, north Ayrshire and Douglas areas. Statistical studies of sections between the Calmy Limestone and Plean No. 1 Limestone in the Kincardine Basin suggest that sand entered the basin from the north-west, north and east, transported by more than one river.

The thickness of the group as a whole is reduced over the Burntisland anticline to about 210 m but this increases again further east in Fife and Midlothian. It is about 500 m thick near Leven in east Fife. In Midlothian the maximum thickness of about 330 m occurs at the west margin of the syncline, around Loanhead and Gilmerton. From there it diminishes towards the south and east.

From the Kincardine Basin there is a general thinning to the south and south-west. The thinning, however, is interrupted by a narrow east-north-easterly basin on the south side of the Campsie Fault, near Kirkintilloch, where a thickness of 405 m has accumulated in the centre of the basin.

In the Glasgow area, the complete succession is about 300 m thick and at several horizons above the Calmy Limestone coarse channel sandstones are present. Further thinning occurs south-westwards, into north Ayrshire, where the thickness is reduced to about 100 m. The latter represents little more than half of the complete sequence as it is known in the Glasgow area. A large proportion of the succession above the equivalent of the Calmy Limestone was removed by erosion prior to deposition of the Passage Group.

The thickness of the group is further reduced abruptly across the line of the Dusk Water Fault from 100 m in the Dalry area to 40 m near Kilwinning. Most of the succession above the Lower Linn Limestone (= Orchard) has been eroded prior to Passage Group sedimentation. In addition, the Index Limestone is locally absent due to erosion by an overlying channel sandstone.

Further attenuation occurs southwards across the Inchgotrick Fault where the thickness below the unconformity is about 8 m. The group recovers towards the south and east on the south side of both faults, so that in the Kilmarnock area about 80 m of strata remain and near Sorn the lower part of the succession, including possibly the Orchard Limestone, is preserved and is about 36 m thick. The correlation of the limestones above the Index Limestone in the area south of the Dusk Water and Inchgotrick faults is uncertain. It is probable that no deposition took place in Upper Limestone Group times in the area south of Ayr and that the group may be represented in places by a marginal sandstone facies.

The succession increases in thickness again on the south side of the Kerse Loch Fault where it is about 210 m thick and the Index, Lyoncross, Orchard and Calmy limestones are present. The number of limestones decreases and the overall thickness of the group diminishes towards the Southern Upland Fault.

The succession at Douglas shows the same pattern of variation in thickness as the Limestone Coal Group. The maximum thickness of 320 m occurs in the north-east part of the area and there is a north-easterly trending zone of minimum thickness between Douglas and Coalburn where it is only 150 m thick.

The Castlecary Limestone is not present in the Douglas area, but the Plean Limestones are represented. The top of the group is taken at the unconformity at the base of the Passage Group which transgresses from a horizon above the Plean Limestones to one just above the Calmy Limestone.

The marine horizons tend to be thicker in the outcrops in the southern part of the Midland Valley. The Index Limestone and particularly the Calmy Limestone with thin associated marine shales are well developed in Douglas and Muirkirk, and in the Dalmellington area. The local equivalent of the Calmy Limestone is about 24 m thick and is characterised by the presence of corals.

Contemporaneous volcanic rocks

Contemporaneous volcanic rocks in the Upper Limestone group are limited to parts of Fife and West Lothian and various occurrences in the Dalry area of north Ayrshire. In the Bo'ness to Bathgate area basaltic lavas and tuffs occupy much of the sequence below the Orchard Limestone. In Fife basaltic tuffs occur at several levels in the succession and lavas occur less commonly.

Passage Group

The uppermost division of the Namurian in the region, the Passage Group, includes the strata between the Castlecary Limestone and the base of the Westphalian (Coal Measures). In those areas where the Castlecary Limestone is missing the base of the group is drawn at the top of recognisable Upper Limestone Group beds and is therefore poorly defined in some areas. The top of the group cannot be accurately drawn in the region as the definitive base of the Westphalian, the Subcrenatum Marine Band has not been recorded in Scotland. The top of the group is taken at the locally defined base of the Coal Measures in the various areas.

The lower part of the group is assigned to the Arnsbergian (E_2) Stage but the later Namurian stages are poorly developed or absent. There is no diagnostic evidence for beds of Chokierian (H_1) and Alportian (H_2) age but miospores indicate that some deposits of the Kinderscoutian (R_1), Marsdenian (R_2) and Yeadonian (G_1) stages are present.

The rocks consist mainly of sandstones and thick beds of clayrock. Thin coals and beds of marine mudstone also occur as minor components of the succession. The clayrocks are of economic importance but the coals are not usually of workable thickness except in the Westfield area of Fife.

Lithology

The rhythmic character of the sediments of earlier groups persists into the Passage Group, but it tends to be obscured by the predominance of thick sandstone beds, many of which have erosive bases. The occurrence of numerous non-sequences which stem from penecontemporaneous erosion has resulted in the impersistence of even well-developed marine horizons, and episodes of coal-formation were short-lived.

The predominant rock-type is sandstone which is commonly medium- or coarse-grained, white, pale yellow or grey in colour and locally reddened. The sandstones frequently contain pebbly bands which in the lower part of the group may include fragments of ironstone, coal or oil-shale. The constituents of the sandstones are predominantly quartzose and they are classified as orthoquartzites and protoquartzites with a few subgreywackes.

The clayrocks are unbedded or poorly bedded. Some contain roots and are therefore seatclays, but in others there is no sign of roots. A red, purple or yellow mottling locally replaces the more usual grey colour and it has been suggested that this may be due to partial oxidation during periods when the water-table was temporarily lowered. Some of the clayrocks are refractory and are a valuable raw material. They are grouped into the Lower and Upper Fireclays and it is the former which is the most sought after.

There are several marine bands in the succession (Figure 24). They are usually thin, consisting of shale with marine fossils and, in the lower part of the succession particularly, contain thin bands of shelly limestone. The most widespread limestone is the Roman Cement or No. 2 Marine Band which occurs in the lower part of the sequence.

Coal seams and ironstones are present in the succession and are quite numerous, but they are thin and few have been worked.

Correlation

The marine bands in the Passage Group are best developed in the Kincardine Basin where the succession is thickest. However, the correlation of marine bands is hardly possible from one area to another. They tend to be impersistent due to minor non-sequences and there is no locality at which all known marine bands are present.

The marine bands were originally numbered from 1 to 3 in ascending order, but subsequently others were found and the numbering scheme has had to be modified. Currently there are Nos. 0, 1 and 2 Marine Bands and Nos. 3, 5 and

6 Marine Band Groups (Figure 24). Up to 16 or 17 individual marine beds are now known but many occur only locally.

Lateral variation

The thickest development of the Passage Group, about 335 m, occurs in the Kincardine Basin. Sixteen or seventeen individual marine bands are present but they tend to be reduced in number and in variety of fauna where the sequence is thinner. Many are also cut out by local non-sequences.

The two principal components of the succession, the sandstones and the clayrocks are subject to marked lateral variation in thickness and lithology.

The commercially important 'Lower Fireclays' lie between No. 2 Marine Band and No. 3 Marine Band Group and individual bands can be up to 18 m thick in places. It is not possible to correlate bands of fireclay even over fairly short distances. The 'Upper Fireclays' overlie No. 6 Marine Band Group.

Two coals seams are fairly persistent throughout the area. The Netherwood Coal occurs within the No. 3 Marine Band Group and the Bowhousebog Coal occurs within the 'Upper Fireclays'. The top of the group is taken at the base of the Lowstone Marine Band.

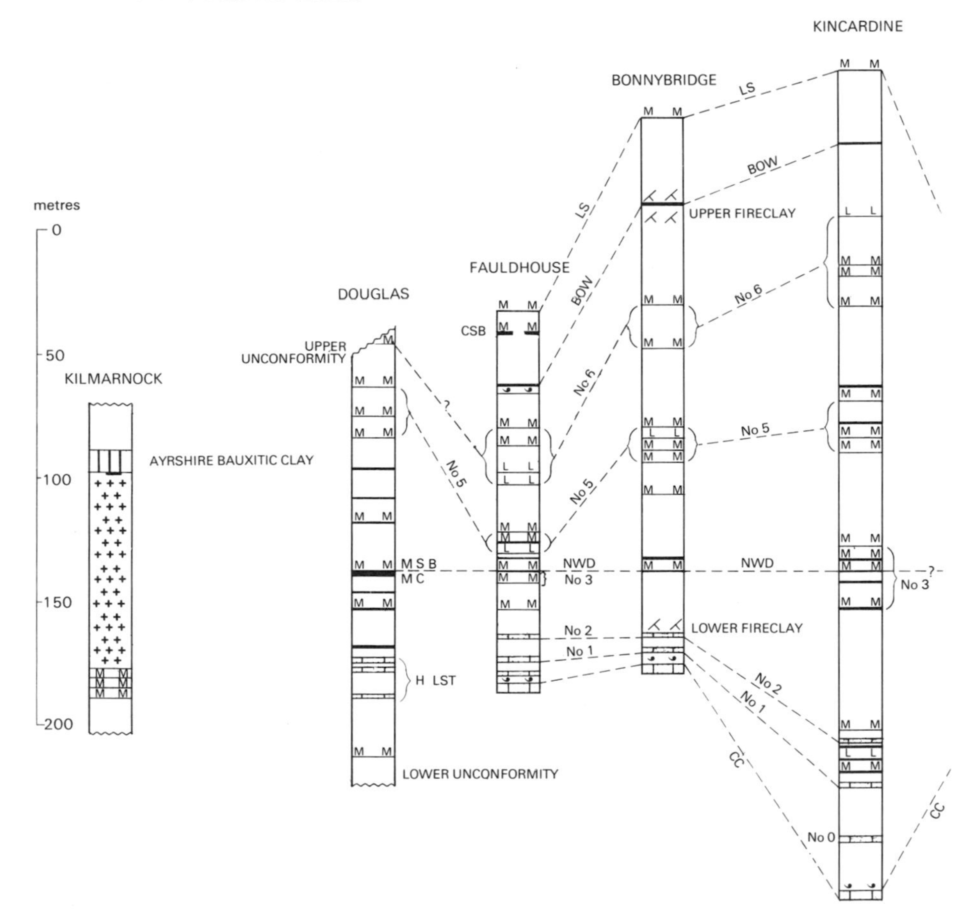

Figure 24 Comparative generalised vertical sections of the Passage Group

In Midlothian the Passage Group, formerly known as the Roslin Sandstone, is up to 240 m thick. It is predominantly arenaceous and is locally reddened. The number of marine bands is greatly reduced compared with the Kincardine Basin succession and correlation with that area is not possible.

The top of the group is lithologically determined and is taken at the base of the Seven Foot Coal, the local base of the Coal Measures.

In the Douglas outlier the typical Passage Group lithologies of sandstones and clayrocks are developed and locally there are unusually thick coals. The base of the sequence is an unconformity which rests on various levels in the upper part of the Upper Limestone Group. The top of the group is placed at the base of the Porteous Band which is a marine horizon marking the local base of the Coal Measures.

An unconformity in the upper part of the group cuts out part of the sequence (Figure 25). It has least effect in the area around Happendon, but progessively more of the upper part is cut out to the north, west and south. In the area south-east of the Kennox Fault the group is cut out completely and rocks of Coal Measures age rest unconformably on rocks of the Limestone

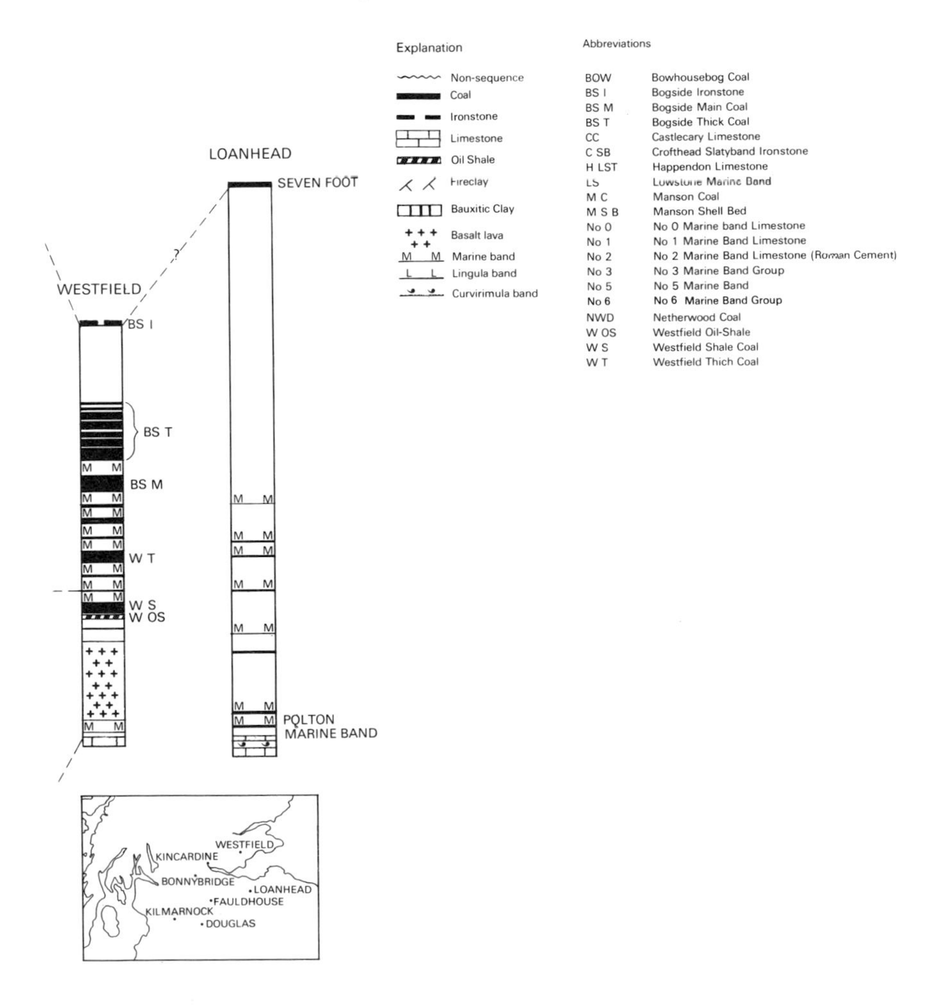

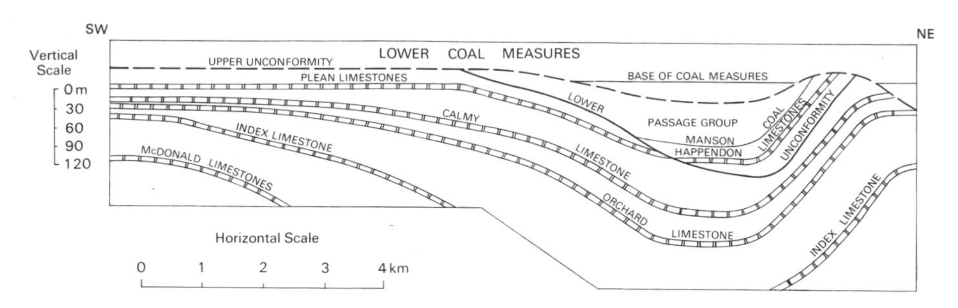

Figure 25 Diagrammatic section showing relation of strata above and below the lower and upper unconformities in the Carboniferous sequence of Douglas

Coal Group and Lower Limestone Group. The maximum residual thickness is about 200 m.

There are several coal seams but they are only a few centimetres thick. Mottled red, yellow and grey clayrocks occur in the upper half of the sequence.

A group of three thin limestones called the Happendon Limestones occur in the lower part of the sequence and these may correlate with the Nos. 0, 1 and 2 Marine Bands of the Kincardine Basin. The Manson Shell Bed which lies in the middle of the succession is possibly equivalent to the No. 3 Marine Band Group and consequently the underlying Manson Coal could correlate with the Netherwood Coal. The Manson Coal is as much as 6 m thick in places.

The development at Westfield, in Fife, is extraordinary. The Passage Group rocks occur in a synclinal outlier lying between the Kincardine Basin and the east Fife–Midlothian Basin and it is thinner than the sequence in either of these two areas. The succession can be subdivided into three parts. The lower part rests on the Castlecary Limestone and consists of sandstones and a lava flow. The middle subdivision, called the Boglochty Beds, contains a number of very thick coals. The thickness varies from 30 to 150 m and one-third of that thickness consists of coal. In addition to coal there are two oil-shale seams, the Westfield Shale and the Canneloid Shale. Marine bands occur at several levels in the Boglochty Beds and miospore evidence suggests a correlation of one of the bands with the No. 3 Marine Band Group in the Kincardine Basin. The upper subdivision consists mainly of sandstone.

The top of the Passage Group is taken at the base of the Bogside Ironstone and Coal which is an horizon that can be correlated with other areas. However, the evidence of miospores indicates that the Namurian/Westphalian boundary falls near the top of the Boglochty Beds.

At Westfield there has been a large opencast mine, over 200 m deep from which the coals of the Boglochty Beds, amongst others, have been extracted.

In Ayrshire, north of the Kerse Loch Fault, the Passage Group rocks consist of a thin and variable lower sedimentary subdivision, the Passage Group Volcanic Formation and the Ayrshire Bauxitic Clay.

The lower sedimentary subdivision consists of sandstones and clayrocks with bands of marine shale. The formation is 30 m thick at its maximum and rests unconformably on various horizons in the Upper Limestone Group.

The Passage Group Volcanic Formation consists of a pile of basalt lavas with minor intercalations of sandstone and clayrocks. The formation is thickest in the Troon area where it measures about 150 m.

The Ayrshire Bauxitic Clay varies in thickness from about 1.5 to 9 m and is the product of decomposition and leaching of the underlying lavas. The rock consists mainly of kaolinite with minor amounts of the bauxite minerals boehmite and diaspore, but varying amounts of iron and silica limit the economic value of the deposit. It has been worked at several localities for the chemical industries. The top of the Passage Group is the top of the Ayrshire Bauxitic Clay.

In south Ayrshire the group is poorly known but in the New Cumnock area about 80 m of strata have been assigned to it. The beds consist mainly of sandstones and clayrocks but two thin limestones are present.

9. Westphalian

The Westphalian succession in the Midland Valley consists of the Lower, Middle and Upper Coal Measures. During the Lower and Middle Coal Measures sedimentary conditions were similar to those in the Limestone Coal Group of the Namurian Series, but the Upper Coal Measures are more arenaceous and show evidence of increasing aridity in the upper part. The rocks are mudstones, sandstones and siltstones with seams of coal and seatclay and rarely of ironstone deposited in cyclic sequence. A few thin marine bands occur which are important for correlation. Coal seams are well developed in the Lower and Middle Coal Measures and up to 20 seams have been worked extensively. The sediments in the Upper Coal Measures are secondarily reddened and coal seams are mostly destroyed by oxidation.

The strata were deposited in a fluvio-deltaic environment in which the marine influence was restricted to a few short-lived intervals. Although subsidence was considerable in order to receive the large amount of sediment brought into the area, sedimentation kept pace to the extent that the water depth was never great and the depositional surface was always close to sea level. However, the degree of subsidence varied greatly from place to place with the contemporaneous formation of local basins and differential subsidence across lines of faulting.

The rock sequence consists of the repetition, in whole or in part, of a sedimentary cycle which commonly has a mudstone at the base and a root-bed and coal at the top. Beds of sandstone, siltstone and mudstone may appear in the cycle in a number of combinations below the root bed. Marine horizons are rare, but strata with non-marine bivalves or 'mussels' are not uncommon. The cycles range in thickness up to about 30 m but average 10 m. There is considerable variety, within these limits, between one cycle and another and laterally within a single cycle.

Apart from fragmentary plant material, the most abundant fossils present are the non-marine bivalves (mussels) which in conjunction with the marine bands, form the basis of correlation both within and between coalfields. The Vanderbeckei and Aegiranum marine bands contain species unique to the individual horizons and to the equivalent bands elsewhere in western Europe. This suggests that the depositional facies was uniform over a very large area in Westphalian times. The presence of these bands in Scotland allows the recognition of Westphalian A, B and C of the western European classification (Table 4).

In the Upper Coal Measures evidence of marine conditions is scarce and confined to the lowermost parts of the sequence. Mussels become scarce compared with their occurrence in the Middle Coal Measures.

The base of the Lower Coal Measures is drawn at a locally convenient horizon in each coalfield area since the Subcrenatum Marine Band, which marks the base of the Westphalian, has not been recorded in Scotland. Its horizon probably lies within the upper part of the Passage Group. The base of the Middle Coal Measures is defined as the base of the Vanderbeckei

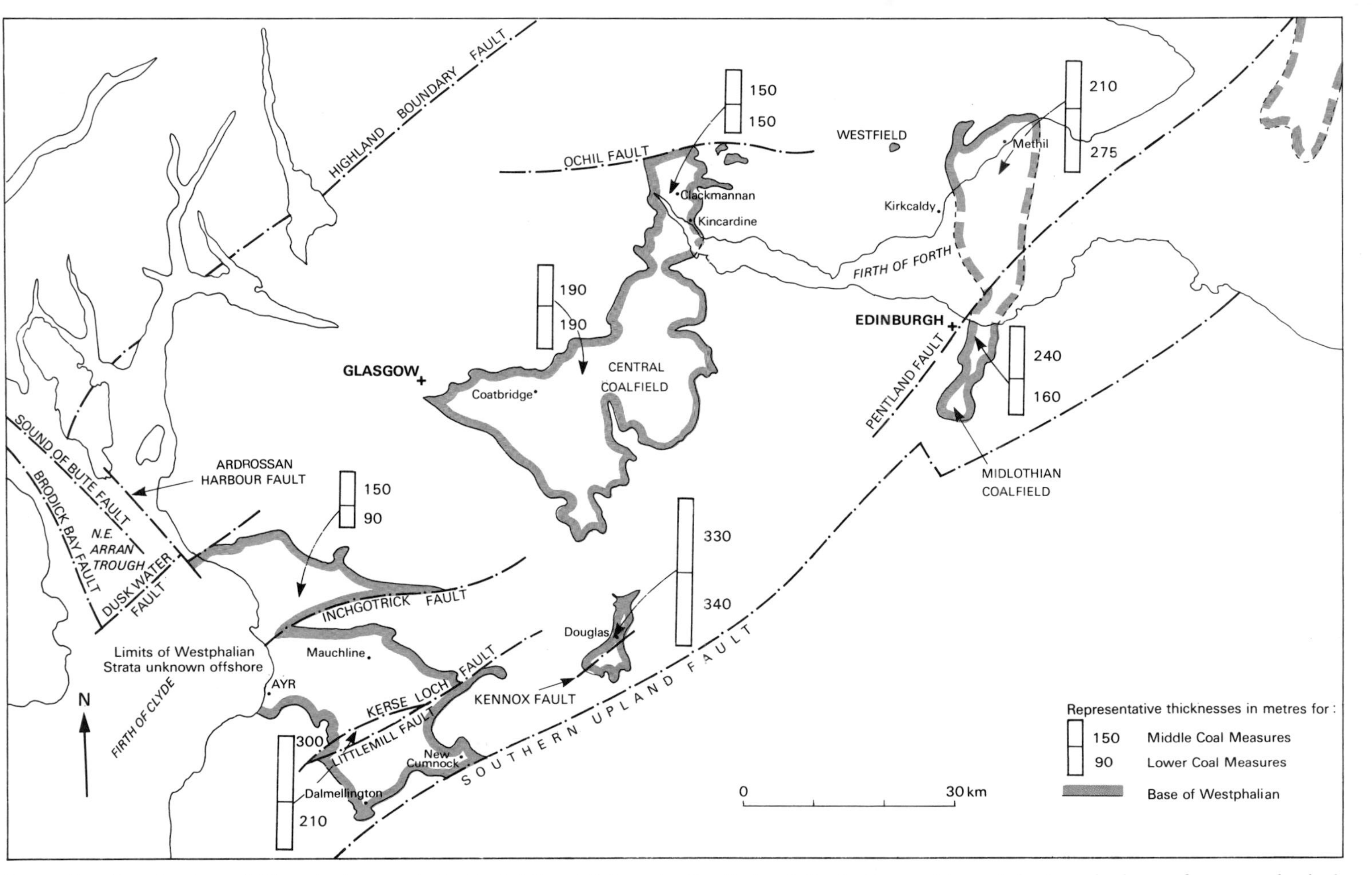

Figure 26 Distribution of Westphalian strata in the Midland Valley. In the Firth of Forth the position of the base is drawn from geophysical data and some boreholes and is only approximate. The limits in the Firth of Clyde are unknown

(Queenslie) Marine Band and the base of Upper Coal Measures is drawn at the top of the Aegiranum (Skipsey's) Marine Band.

The distribution of the Westphalian outcrops is shown on Figure 26 which shows that the sequence is now only preserved in a few separate areas. The outcrops shown offshore in the Firth of Clyde and the Firth of Forth are known from geophysical data and a few boreholes. Knowledge of the succession in these areas is meagre. The generalised successions in the various coalfields are shown on Figure 27 together with the stratigraphical classification and the names of some of the principal coals.

Lower and Middle Coal Measures

The Lower and Middle Coal Measures were formerly called the Productive Coal Measures to distinguish them from the overlying Upper Coal (Barren Red) Measures. The limits of the two divisions have been given above and the major features of the stratigraphy and classification based on marine bands and non-marine bivalves are shown on Figure 27.

The Westphalian of the British Isles has been subdivided into a series of chronozones which are based on the distribution of the numerous species of non-marine bivalves. In the Lower and Middle Coal Measures (Westphalian A and B) the zones are, in ascending order, the Lenisulcata, Communis, Modiolaris and Lower Similis-Pulchra chronozones. In the Midland Valley only the upper part of the Lenisulcata Chronozone can be recognised from the mussels, and this only in parts of Ayrshire and the Central Coalfield, but typical faunas of the other chronozones are present over the region.

While mussels are mainly useful for identifying ranges of strata, marine bands are more useful for the identification and correlation of individual beds. A small group of up to three bands is present near the top of the Lenisulcata Chronozone but they normally only contain *Lingula*. The Vanderbeckei Marine Band lies in the middle of the Modiolaris Chronozone. It consists of mudstone and contains a varied fauna of bivalves and goniatites but yields only *Lingula* in some areas.

A group of up to four bands occurs in the Lower Similis-Pulchra Chronozone which usually only contains *Lingula* but bivalves are present in a few areas. The Aegiranum Marine Band marks the top of the Middle Coal Measures and of Westphalian B, and contains the most varied fauna of all of the bands. The bed is developed as a limestone in some areas and carries a rich fauna including calcareous brachiopods, goniatites and nautiloids.

Numerous thick coals occur throughout the succession but the bulk of the worked seams are in the Communis, the upper part of the Modiolaris and the lower part of the Lower Similis-Pulchra chronozones. Many of them split into two or more seams when traced laterally and correlation is achieved by a comparison of lithological sequences and from faunal evidence.

In most parts of the region the Lower Coal Measures rest conformably on the Passage Group. The main exception is in north Ayrshire where the Passage Group contains volcanic rocks, and mussels of Communis age occur 10 m above the lavas.

Lithology

The rocks consist of white or grey sandstones, grey or dark grey siltstones and mudstones with ironstones, and numerous coals and seatclays. The sandstones are locally quite coarse especially in the lower part of the sequence where they form the dominant lithology in a rather variable development. Bands of nodules of clayband ironstone are common and blackband ironstone was formerly worked in several areas. Some of the major coal seams are persistent within a coalfield although splitting laterally into two or more leaves is not uncommon.

Lateral variation

Originally the Lower and Middle Coal Measures probably extended over the whole of the Midland Valley, although evidence is lacking north of the Ochil Fault, and they overstepped on to the relatively positive Southern Uplands. In the eroded remnants of the outcrop the greatest thickness is preserved in the relatively confined areas between parallel faults associated with the Southern Upland Fault in the Douglas area and in the Littlemill area of south Ayrshire (Figure 26). The zone of attenuation of strata, approximately coincident with the Burntisland Anticline, which affects relative thicknesses of older Carboniferous groups is much less apparent in the Lower and Middle Coal Measures. Comparative vertical sections of the Lower and Middle Coal Measures are given in Figure 27.

In the Kincardine area the strata are contained in a N–S-trending syncline which plunges northwards north of Clackmannan and southwards south of Clackmannan. The outcrop is bounded to the north by the Ochil Fault against which the strata are upturned.

The base of the Coal Measures in this area and throughout the Central Coalfield is taken at the Lowstone Marine Band which is fairly widespread although the correlation becomes more tentative in the Glasgow area. The Vanderbeckei and Aegiranum marine bands occur throughout the area except in the Kincardine area where beds of the latter age have been removed. All four non-marine bivalve chronozones are represented.

The Lower Coal Measures succession is thickest in the axial part of the syncline, around Clackmannan, where it is about 150 m thick. The Middle Coal Measures are also about 150 m thick. Attenuation is considerable towards the north, but less so in other directions, repeating the pattern of thickness variation that occurred during deposition of the Namurian. Further south in the Central Coalfield the Lower Coal Measures and the Middle Coal Measures are each about 190 m thick. The Lower Coal Measures thin westwards to about 100 m in the Glasgow area. Comparative thicknesses have to be regarded with caution in view of the uncertainty of the correlation of the horizons used as the base.

In the Clackmannan and Coatbridge areas the upper part of the Middle Coal Measures consists of thick red sandstone rather than the usual grey mudstone, pale sandstone and coals.

In the small outlier at Westfield, in Fife, the Lower Coal Measures are about 160 m thick and the base is placed at the Bogside Ironstone and Coal. The Middle Coal Measures are about 200 m thick. In east Fife the Lower Coal Measures are up to 275 m thick, but offshore in the Methil area they consist mostly of tuffaceous rocks. The Middle Coal Measures are up to 210 m thick.

In the Midlothian Syncline the base of the Coal Measures is taken at the Seven Foot Coal in the north and at the Melville Group of coals in the south.

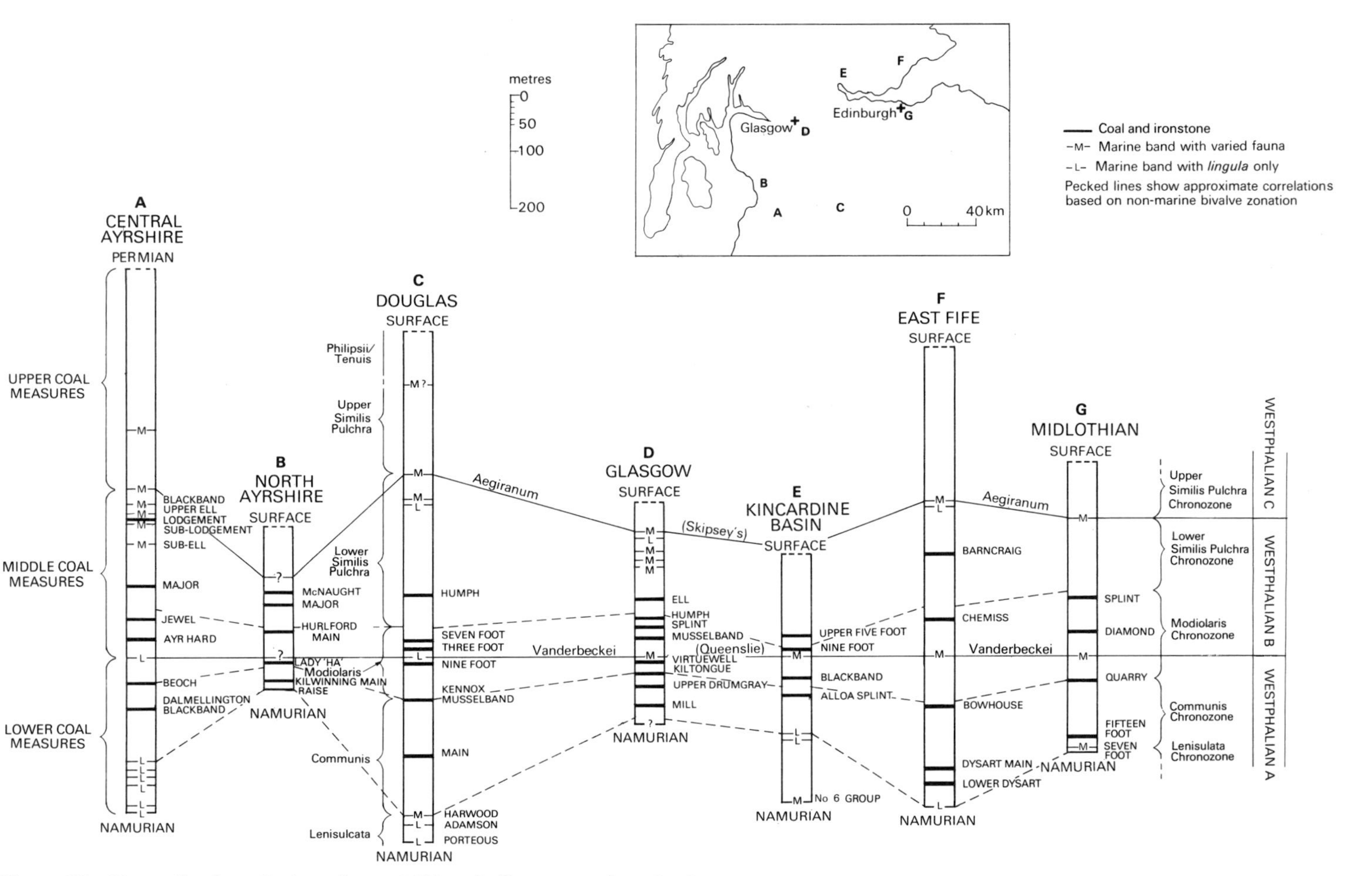

Figure 27 Generalised vertical sections of Westphalian successions in the Midland Valley showing classification and zonation

The Lower Coal Measures are about 160 m thick and the Middle Coal Measures measure about 240 m.

The non-marine bivalve faunas found in Midlothian are relativelv poor. but the Communis, Modiolaris and Lower Similis-Pulchra chronozones are all represented.

In the Douglas area the Lower Coal Measures are up to 340 m thick and the base is placed at the Porteous Band, which is a marine band containing *Lingula* only. The Middle Coal Measures are about 330 m thick. There is little variation in thickness of the strata.

All four non-marine bivalve zones are present although the evidence for the Lenisulcata Chronozone is slight. Three marine bands, including the basal Porteous Band occur in the Lower Coal Measures and all are included in the Lenisulcata Chronozone. Both the Vanderbeckei and Aegiranum Marine bands are present and two other marine bands occur, one with *Lingula* alone, in the upper part of the Middle Coal Measures.

An unconformity underlies the greater part of the Coal Measures outcrop. Only in the north of the outlier is there continuity from the Passage Group into the Coal Measures. Elsewhere the basal beds of the Coal Measures rest on rocks ranging in age from Lower Limestone Group to Passage Group. The precise age of the Coal Measures strata overlying the unconformity is unknown.

In Ayrshire, the Lower and Middle Coal Measures show a pattern of thickness variation which is similar to the variation seen in the Namurian but it occurs in a lesser degree.

The succession is thickest in the Dalmellington and New Cumnock areas and in the Littlemill Trough between the Kerse Loch Fault and the Littlemill Fault. The Lower Coal Measures are up to 210 m thick and the Middle Coal Measures are about 300 m thick. The thickness is reduced abruptly northwards across the Kerse Loch Fault and there is further attenuation north-westwards towards the Inchgotrick Fault.

Comparison of thickness is dependent on the equivalence of the horizons used as the base of the Coal Measures, and in Ayrshire different horizons are used in different areas. South of the Kerse Loch Fault the Passage Group/Lower Coal Measures boundary is not well defined and it is probable that it should be drawn lower than it is shown on Figure 27, column A. In east Ayrshire the base is taken at the second *Lingula* band below the Pathhead Thick Coal.

The reduction in thickness northwards and westwards is due in part to an unconformity at the base of the Coal Measures in the area to the north of the Kerse Loch Fault. In this area the Coal Measures are resting on Passage Group lavas and the basal beds become progressively younger northwards and westwards so that on the south side of the Inchgotrick Fault the local base is only a few metres below the Vanderbeckei Marine Band. The succession increases in thickness again north of the fault where the lowest beds belong to the Communis Chronozone.

Musselbands in the succession south of the Kerse Loch Fault indicate that the Lenisulcata Chronozone is present, but its occurrence on the north side of the fault has not been proved.

The Vanderbeckei Marine Band has been found in most parts of the outcrop except north Ayrshire. In this area the horizon of the Shale Coal is taken as its equivalent.

The Aegiranum Marine Band occurs in south Ayrshire and parts of central and north Ayrshire, but may be absent locally in the northern part of the outcrop.

Four other marine bands occur in the upper part of the Middle Coal Measures. They have their optimum development in the Littlemill Trough but elsewhere they are not all represented.

Contemporaneous igneous activity

Evidence of contemporaneous igneous activity in the Lower Coal Measures was found in boreholes in the Firth of Forth, off Methil and Kirkcaldy. In the area east of Methil the Lower Coal Measures sediments are replaced by tuffaceous rocks and a basalt lava flow. South-east of Kirkcaldy thin tuff bands occur in the Lower Coal Measures succession.

Upper Coal Measures

The Upper Coal Measures include the strata from the top of the Aegiranum Marine Band up to the unconformity at the base of the Permian. They are placed in Westphalian C and D in the Western European classification. The strata contain few mussels compared with the Middle Coal Measures but there is evidence that the Upper Similis-Pulchra, Phillipsii and Tenuis chronozones of the non-marine bivalve zonation are present in the region. There is insufficient evidence to separate the last two chronozones.

There is no evidence of marine conditions later than the Upper-Similis Pulchra Chronozone. In that zone in England there are three marine bands but in the Midland Valley the bands are poorly developed and only the uppermost one has been identified on faunal grounds in Ayrshire and part of the Central Coalfield. There is also a record of one of the two lower bands in Ayrshire.

The strata are stained red in most areas and were formerly called the Barren Red Measures because the processes of oxidation which produced the red coloration also caused the alteration or destruction of most of the coal seams. However, the lower limit of reddening is an irregular horizon which, in places, extends well below the Aegiranum Marine Band but elsewhere fails to penetrate to that level.

Lithology

The sediments of the Upper Coal Measures in the lower part at least are essentially a sequence of fluvio-deltaic coal-bearing cycles, similar to those of the Lower and Middle Coal Measures. In many places they have been chemically, and to some extent physically, altered by processes of oxidation.

Unaltered, predominantly grey-coloured Upper Coal Measures strata occur in parts of Ayrshire and the Douglas area. The rocks consist of sandstones, locally thick and coarse in south Ayrshire, grey mudstones, seatclays, thin coal seams and nodular clayband ironstones. The coals are poorly developed and they tend to be impersistent laterally. In Ayrshire and the Central Coalfield thin bands of compact cream-coloured limestone occur and are fairly persistent. The limestones are normally less than 0.3 m thick and contain *Spirorbis*, ostracods, estheriids and fish fragments. The rocks are not well known because of poor exposure and a lack of borehole and mining information.

The alteration affects all rock-types to some degree. The colour of the rocks is changed mainly to red, but also in places to green, lilac or yellow. The reddening is most intense in the upper part of the sequence where about 300 m are affected, but partial oxidation penetrated deeper, facilitated in some instances by fault planes and more permeable layers. The red coloration is due to the alteration of various iron minerals to hematite.

The colour change is accompanied by a textural change in the case of the fine-grained sediments. Traces of bedding in rocks which were originally mudstones or silty mudstones become less distinct and such rocks have been described as clayrocks or marls. The bedding tends to be further obscured by an apparent brecciation which is in fact merely a colour pattern reflecting contrasting concentrations of hematite and other iron-bearing minerals. Carbonaceous material, including plant debris and rootlets in seatclays has been destroyed but impressions may be left, often with a green coloration.

Coals are altered either by transition to limestone (Mykura, 1960) or removed altogether with only a thin layer of fissile clay or ironstone remaining in the coal position. The complete removal of the coal occurs in the upper part of the reddened succession and the replacement by limestone occurs in the lower part where oxidation is not so far advanced. It is not clear whether this indicates two stages in the oxidation process or that carbonate has been introduced perhaps from the weathering of the Permian lavas.

The reddening of the strata is mainly secondary in origin but it is possible that primary red beds are also present. It is clear that where coal positions can be identified in reddened strata, the reddening must have been secondary. In Fife thick beds of clayrock, within reddened strata containing coal positions, have been described as primary red beds.

Lateral variation

The thickness of the Upper Coal Measures is everywhere a residual one as the Permian lavas unconformably overlie the Upper Coal Measures in Ayrshire. The maximum thickness occurs in the Mauchline basin in Ayrshire where there are about 465 m of strata. All the Upper Coal Measures north of the Kerse Loch Fault are reddened, but in the Dalmellington area and in the Littlemill Trough the lower limit of reddening is 70 to 80 m above the Aegiranum Marine Band. The occurrence of desiccation cracks and calcareous concretions in the uppermost part of the sequence is evidence of increasing aridity.

In the Douglas area there are about 270 m of strata, mostly grey with only local reddening. Comparable thicknesses remain in Midlothian, Fife and the Central Coalfield, but the strata are mostly reddened.

10. Carboniferous palaeontology

The abundant Scottish Carboniferous fauna and flora have been studied in detail for many years and the accumulated information is of prime importance in two respects. Firstly, it greatly assists comparison and correlation of stratal sequences in different parts of the region and many individual beds can be recognised by their faunal content. Secondly, the fossils present give an insight into the environment of deposition of the host rocks. The vertical ranges of selected species are shown on Figures 28 and 29.

Classification and zonation

As a result of the cyclic nature of the sedimentation, each successive depositional facies persisted for a relatively short time. The fossil record shows a sequence of colonisations and extinctions of the various organisms suited to the prevalent facies over all or part of the region. Sediments deposited in marine environments form a relatively small part of the total succession and they tend to be concentrated in the Lower and Upper Limestone Groups. Any scheme of zonation based on marine organisms can only be of limited application and has to be complemented by schemes based on non-marine faunas and on microfloras.

Corals form a very small proportion of the marine faunas and are restricted in their lateral distribution so that a zonal scheme based on them is only of very general use. Goniatites are also confined to marine deposits and are seldom of common occurrence in the Scottish Carboniferous. In addition, the genera of goniatites which have proved to be most useful in zoning the upper Dinantian and Namurian of England are of extremely rare occurrence in Scotland. Despite these limitations studies on goniatites have shown that the base of the Namurian lies just below the Top Hosie Limestone and that the base of the Arnsbergian (E_2) Stage of the Namurian is below the Orchard Limestone. In addition the presence of *Anthracoceratites vanderbeckei* in the Vanderbeckei (Queenslie) Marine Band indicates that the base of Westphalian B of the western European classification should be drawn at this horizon.

Non-marine bivalves are found at some horizons throughout the Dinantian and Namurian strata. In the Westphalian, however, numerous species of several genera occur in profusion at many horizons. A zonation based on these forms in the English and Welsh coalfields is also applicable to the Scottish Coal Measures.

Plant miospores are exceedingly small and occur in myriads at many horizons in the succession. Early work to use them for zonal purposes was restricted to microfloras obtained from coals and was not entirely successful. Subsequent research resulted in a division of the Dinantian of the east of Scotland into five concurrent range zones based on miospores extracted from marine and non-marine mudstones and siltstones. This zonation has yet to be extended to cover the Namurian and Westphalian of the Midland Valley.

At present no single group of fossils can be used to subdivide the Scottish succession and the lithostratigraphical classification is the most suitable one for descriptive purposes.

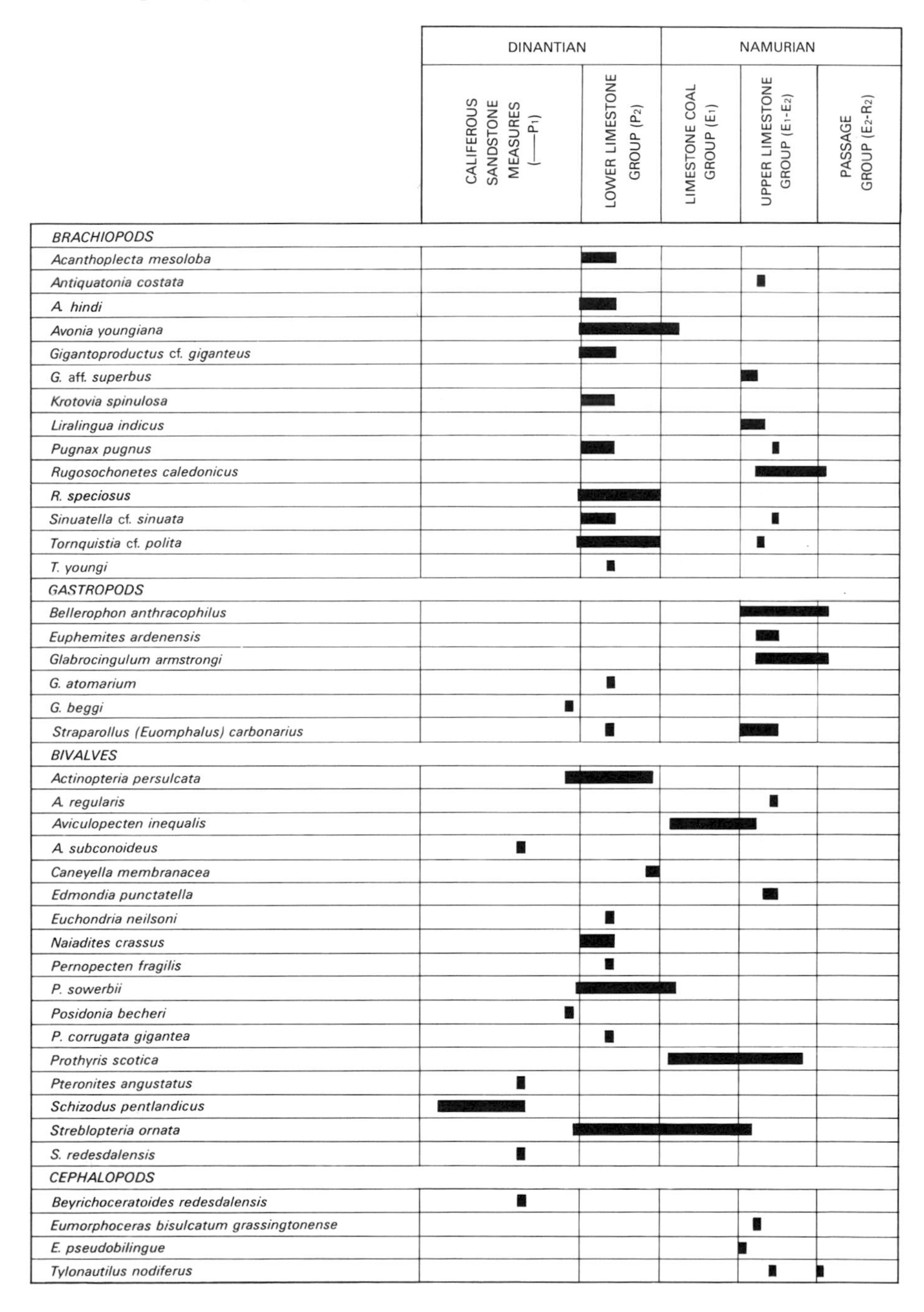

Figure 28 Distribution of some Dinantian and Namurian fossils in the Midland Valley

Faunas

Calciferous Sandstone Measures

The strata of the Calciferous Sandstone Measures, the lower and major division of the Scottish Dinantian, are remarkable for the degree of lateral variation they show when traced across central Scotland. This phenomenon is reflected in the faunas except in the uppermost beds of the division.

The fossils suggest that the major part of the sediments was deposited in non-marine conditions. The most commonly occurring forms are the calcareous-tubed worms *Serpula* and *Spirorbis*, the bivalves *Paracarbonicola* and *Naiadites*, mainly in the lower part succeeded by *Curvirimula* in the higher beds, and local accumulations of arthropods, particularly ostracods, and fish remains.

The first indication of the transgressing early Carboniferous sea is shown by the occurrence of the bivalve *Modiolus latus* in the basal cementstone facies in several areas. This species occurs in marine assemblages but was also able to thrive in conditions, perhaps brackish or hypersaline, which were unsuitable for other marine forms. The earliest varied marine assemblages are probably those associated with the limestone at Wormistone and Randerston in east Fife. Bryozoa, the gastropod *Bellerophon randerstonensis* and bivalves such as *Schizodus pentlandicus* are present. Other marine bands mainly yielding bivalves such as *Leiopteria hendersoni, Sanguinolites clavatus* and *Schizodus pentlandicus* occur in the lower part of the group in east Fife, Midlothian and East Lothian but they probably only represent minor, local marine incursions.

The first major marine episode which can be correlated over a large part of the region is that recorded by the Macgregor Marine Bands. These comprise up to three separate bands which are found in east Fife and the Lothians. They contain a rich marine assemblage of bryozoa, brachiopods and molluscs with *Punctospirifer scabricosta, Pteronites angustatus* and *Streblopteria redesdalensis* the characteristic species. There is evidence that the sea invaded the region from an easterly direction.

It is only in the upper part of the Calciferous Sandstone Measures that marine beds appear in the western part of the region in the Glasgow, Douglas and north Ayrshire areas. The horizons of the Hollybush and Blackbyre limestones of the Glasgow area can be traced with reasonable certainty across the region to East Lothian. These mark the first occasions when marine conditions prevailed across the region and rich faunas of corals, brachiopods and molluscs populated the sea bottom. The bivalve *Actinopteria persulcata* is commonly found in these beds and the Hollybush Limestone is characterised by an abundance of the productoid *Semiplanus* cf. *latissimus*.

Lower Limestone Group

The Lower Limestone Group is the upper division of the Dinantian succession in the Midland Valley and it marks the acme of marine conditions in the Scottish Carboniferous. The limestones and mudstones in the lower part contain rich faunas of corals, bryozoa, brachiopods, molluscs, crinoids and trilobites, many species being confined to the Group. The most distinctive fauna is that of the Neilson Shell Bed which is correlated from north Ayrshire to East Lothian and Fife. Some of the species such as the brachiopod *Tornquistia youngi* and the bivalve *Posidonia corrugata gigantea* appear to be confined to this horizon. The Hosie Limestones and their equivalents in the upper part of the Group are characterised by an abundance of the brachiopods *Eomarginifera, Pleuropugnoides, Productus* and *Schizophoria* and the bivalves *Posidonia corrugata* and *Sanguinolites costellatus. Curvirimula* is the predominant non-marine bivalve in the Group.

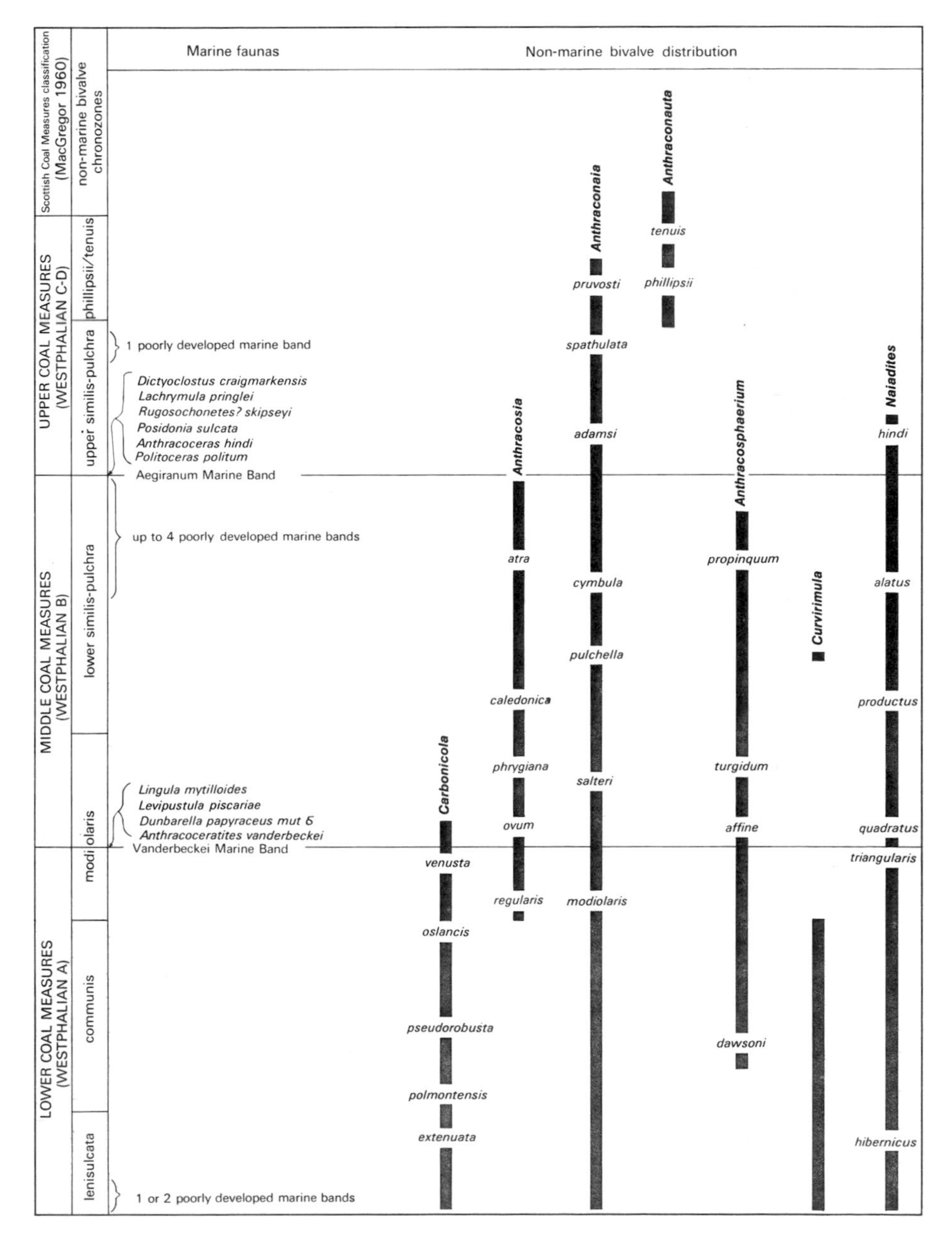

Figure 29 Distribution of stratigraphically important fossils in the Westphalian

Limestone Coal Group

The Limestone Coal Group is the lowest division of the Scottish Namurian succession. The prevailing conditions of deposition were non-marine as is evidenced by the number of coals developed. Only two major marine bands are present over the whole region and these are the Johnstone Shell Bed and the Black Metals Marine Band. Both marine bands are in the lower half of the Group and the faunas are mainly of brachiopod and mollusc species which were present in the region in Lower Limestone Group times. The Group contains numerous bands yielding *Lingula* only, some of which occur over a large area. These are interpreted as partially developed marine incursions in which only *Lingula* was able to thrive in the environment. Non-marine bivalves are represented by *Curvirimula, Naiadites* and *Paracarbonicola*. Fish remains are reasonably common at some horizons and occurrences of amphibia have been recorded from the Group.

Upper Limestone Group

Following the predominant coal-swamp conditions of the Limestone Coal Group the Upper Limestone Group marks the return of several widespread marine episodes with limestone deposition and rich faunas of corals, bryozoa, brachiopods, molluscs and trilobites.

There are consistent differences in the faunas of the Group between those in the Kincardine Basin and Central Coalfield to the west and those of east Fife and Midlothian to the east. An outstanding example of this is the abundance of latissimoid productoids in the Index Limestone in the western area and their rarity at this level in the east whereas in the Orchard Limestone the reverse situation exists. Such differences persist between the faunas of the two areas throughout the Group except in the topmost bed, the Castlecary Limestone. It seems probable that a feature associated with the Burntisland Anticline affected the depositional conditions on either side of the structure which resulted in different facies in the two areas.

Some species are confined to one horizon such as the productoid *Antiquatonia costata* being found only in the Orchard Limestone. The most noteworthy example of the lateral spread of a species is that of the marine bivalve *Edmondia punctatella* just below the Calmy Limestone. It occurs in profusion in a thin band which can be traced from north Ayrshire to west Fife and as far south as Douglas. It is also present in east Fife and Midlothian but is relatively scarce there.

The non-marine bivalves present in the Group are *Curvirimula* and *Naiadites*.

Passage Group

The fossils in the Passage Group record a gradual withdrawal of the sea from the region. There are several marine bands in the lower part of the Group which contain rich faunas of brachiopods and molluscs. The best developed limestone is in No. 2 Marine Band (Roman Cement) which is packed with orthotetoids and *Schizophoria*. Little other than *Lingula* is present in the few fossiliferous bands in the upper part of the Group. The only notable occurrence of non-marine bivalves is of abundant *Curvirimula* in a carbonaceous shale at the base of the Group in the immediate roof of the Castlecary Limestone.

Coal Measures

During Westphalian times when the Coal Measures were deposited, non-marine conditions prevailed for most of the period. Only on two occasions did the sea become established over the whole region. These are marked by the Vanderbeckei (Queenslie) Marine Band and the Aegiranum (Skipsey's) Marine Band. The first named divides the Lower from the Middle Coal Measures and the second is at the base of the Upper Coal Measures. Both bands contain varied marine faunas including brachiopods and molluscs but the Vanderbeckei Marine Band yields only *Lingula* over much of the region and productoids, pectinoids and goniatites are confined to relatively small areas.

Marine bands are also present locally near the base of the Lower Coal Measures and above and below the Aegiranum Band. They are impersistent in lateral development and normally only contain *Lingula* but marine molluscs are present at some localities.

It is the non-marine bivalves, however, which are the principal fossils in the Coal Measures (Figure 29). At numerous horizons, normally in the mudstone roof of a coal, they occur in vast numbers and some of these 'musselbands' can be traced over large areas. A succession of species of *Carbonicola* are the principal forms in the Lower Coal Measures and these are followed in the Middle Coal Measures by species of *Anthracosia*. A marked diminution of these faunas takes place at the Aegiranum Marine Band. Of the six genera present in the Lower and Middle Coal Measures only *Anthraconaia* and *Naiadites* persist into the Upper Coal Measures and are relatively rare in these beds. In the upper part of the Upper Coal Measures the only fossils found are restricted to scarce plant remains and the worm *Spirorbis*.

11. Permian and Triassic

Rocks ascribed to the Permian and Triassic consist mainly of red sandstones and mudstones with basalt lavas at the base and they contain little fossil evidence of their age. They are known collectively as the New Red Sandstone and the term is applied to rocks younger than Carboniferous but older than Jurassic. They have a very restricted outcrop in the Midland Valley. The Permian is present in the Mauchline Basin and in Arran (outside the present region), and also offshore in the Firth of Clyde and the Forth Approaches. Triassic rocks occur only in south Arran and in the Firth of Clyde. The outcrops in the Midland Valley and neighbouring offshore areas are shown in Figure 30. The offshore outcrops are known from geophysical data, some drill samples and by extrapolation from onshore sections in the Mauchline Basin and on Arran.

A climatic change occurred during the late Carboniferous from humid to semi-arid or arid conditions and the change in climate is reflected in the

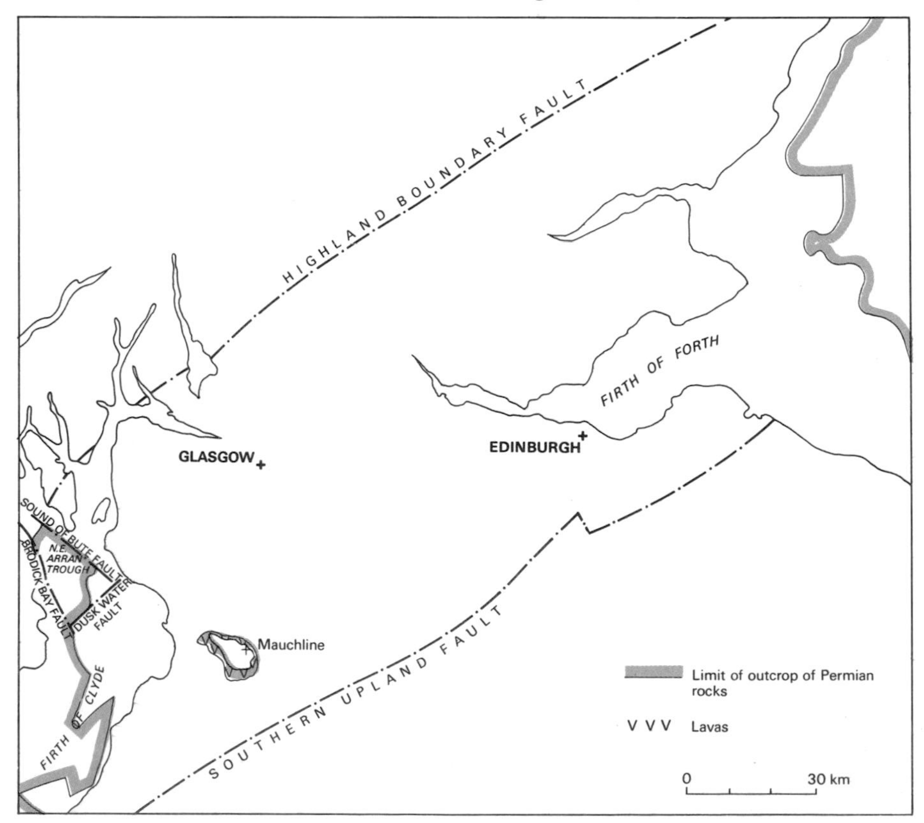

Figure 30 Distribution of Permian rocks in the Midland Valley and adjacent offshore areas

sedimentary sequence. The Upper Coal Measures are fluviodeltaic and fluviatile deposits which show signs of increasing aridity and the Lower Permian sediments are aeolian desert sandstone. There is no apparent angular discordance at the base of the Permian in the Mauchline Basin but there is, however, a distinct lithological break and an unconformity underlies the Permian rocks on Arran. The reddening of the Upper Coal Measures presumably took place in late Carboniferous or Permian times.

The fossil evidence for the age of the oldest New Red Sandstone rocks is weak but it indicates a probable Lower Permian (Autunian) age on the basis of some plant material found in sediments between lava flows near the base of the volcanic succession in the Mauchline Basin (Wagner, 1983). Radiometric dates of 275 to 272 Ma have been obtained by the postassium/argon whole rock method for the Permian lavas near Mauchline.

The rocks of the Mauchline Basin are subdivided into two groups: the Mauchline Volcanic Group at the base is overlain by the Mauchline Sandstone.

Mauchline Sandstone	*m*
Brick-red sandstone characterised by the presence of wind-rounded grains, with cross-bedding of dune type	450+
Mauchline Volcanic Group	
Basaltic lava flows, usually thin, intercalated with beds of agglomerate, tuff, desert sandstone and mudstone	90 to 235
Basal tuffs and sediments	3 to 80

The base of the Permian is placed at the first appearance in the succession of either bright red sandstone with wind-rounded grains or a sediment containing lava fragments. Most commonly the base is a red sandstone with wind-rounded grains, but where the basal beds are thickest they consist of water-lain siltstone and sandstone with lenses of basaltic detritus. Nodular calcareous concretions also occur, which resemble the cornstones of the Upper Devonian, and may be analogous to the caliche formed in soils in present-day arid regions.

Sediments intercalated with the lavas consist of sandstones with pebbly bands of volcanic detritus and thin argillaceous beds. Pyroclastic rocks have a matrix of mud and contain wind-rounded grains of sand.

Both plant remains and the fine-grained, water-lain sediments which occur interbedded with the lavas indicate that the aridity was tempered, at least occasionally, by rainfall.

The Mauchline Sandstone is brick-red to orange-coloured and is the result of the lithification of sand-dunes. The large-scale cross-bedding of the dunes can be seen in the cliffs along the River Ayr near Mauchline. The sandstone is highly quartzose, almost devoid of mica and many of the larger quartz grains are rounded and polished by wind action.

The sandstone was formerly quarried for building stone at Ballochmyle, near Mauchline and was used in many buildings and monuments in the west of Scotland. The quarry has since been used for dumping colliery waste and the face which once displayed fine examples of dune-bedding (Plate 12.1) can no longer be seen.

In the Firth of Clyde the area shown as Permian in Figure 30 is assumed to contain sediments similar to the Mauchline Sandstone. This interpretation is based on geophysical data and a few drill samples.

The Permian rocks in the west of Scotland are terrestrial deposits probably of Lower Permian age and equivalent to the Rotliegendes of the European succession. In the Forth Approaches the Rotliegendes appears to be absent and Upper Permian marine deposits of the Zechstein Sea are deposited on Devonian and Carboniferous rocks. The outcrop is shown on Figure 30 and is based on geophysical data and borehole sampling. The rock samples include gypsum, anhydrite, dolomite, limestone and marl.

Palaeogeography and deposition

In Permian times the region lay within a large continental land mass about 8° north of the equator (see Figure 31). The climate was arid to semi-arid and aeolian sediments indicate that the prevailing wind came from the east.

Deposition of the Lower Permian rocks was by both aeolian and flash flood processes. The Upper Permian in the Forth Approaches on the other hand indicates evaporitic deposition from the hypersaline Zechstein Sea.

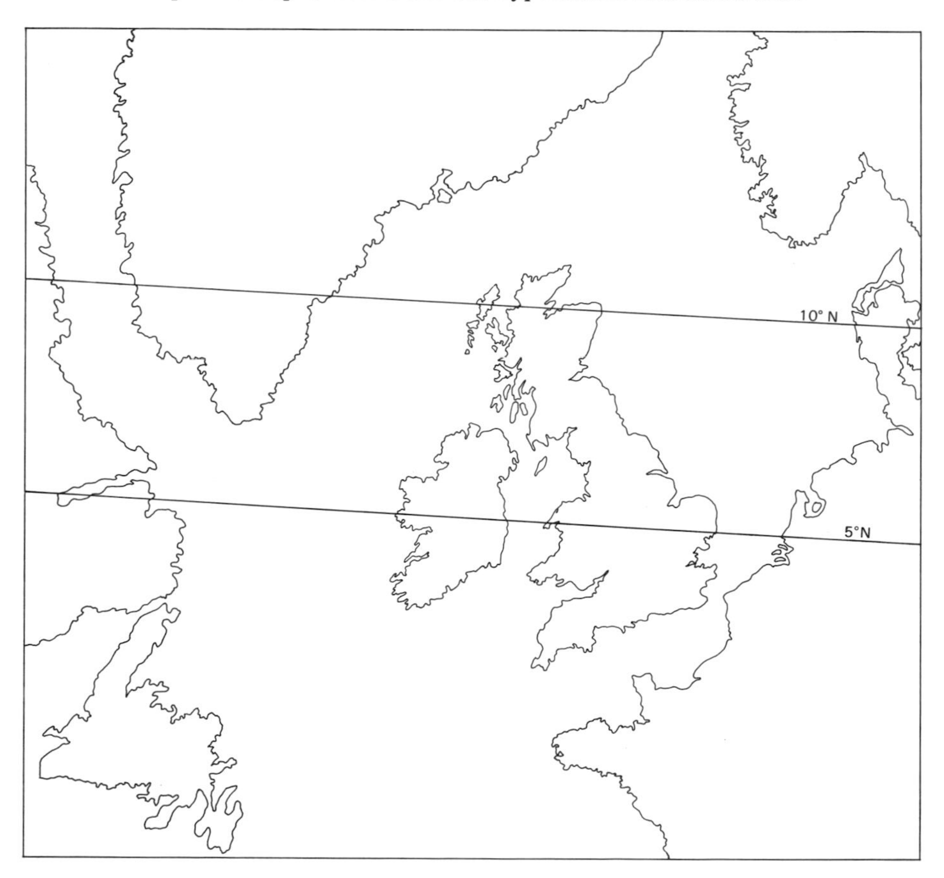

Figure 31 Palaeolatitudes in the Permian (after Smith and others, 1981)

12. Carboniferous and Permian igneous activity

Following a long period of magmatic quiescence during the Middle and Upper Devonian, igneous activity became widespread in the Midland Valley during the early to middle Viséan when thick plateaux of alkali olivine-basalt and related lavas were formed. Thereafter, smaller, localised centres of more explosive pyroclastic activity and/or alkali basalt lava flows erupted somewhere in the Midland Valley throughout almost the entire succession, up to and including the Lower Permian. The later volcanic episodes were accompanied by the intrusion of thick, widespread sill-complexes of various alkali dolerite types, notably during the late Namurian and early Westphalian in the east, and during the Stephanian and Lower Permian in the west. The alkali basalt activity was interrupted during the late Westphalian or early Stephanian by a widespread suite of tholeiitic sills and dykes with no known extrusive equivalents.

The alkali olivine-basalts, dolerites and related differentiates constitute a major continental, alkaline province which has been studied extensively since the early days of igneous geology.

Volcanic activity

Distribution in space and time

Outcrops of Carboniferous and Permian volcanic rocks are shown on Figure 32. Many of the volcanic formations are interbedded with well established, often fossiliferous, stratigraphical successions and it is relatively easy to trace their development. This is particularly true of the late-Viséan and Silesian volcanic sequences and some of the early Viséan outcrops of the eastern Midland Valley. Correlation is less easy in the Viséan of the western Midland Valley where sedimentary intercalations are rare within the thick lava sequences, which replace much of the poorly fossiliferous parts of the Calciferous Sandstone Measures. The principal volcanic developments of the Midland Valley are shown in relation to the established Carboniferous stratigraphy in Figure 33.

Petrography of the lavas

The majority of lavas are basaltic and range from ankaramite to basalt to hawaiite in bulk chemical composition. Lavas of intermediate composition are mainly mugearites, although trachybasalts and trachyandesites are recorded and trachytes, quartz-trachytes and rhyolites are present in some centres. Nepheline-bearing differentiates such as phonolitic trachyte and phonolite are restricted to subvolcanic intrusions.

The basaltic rocks are almost always porphyritic, enabling a classification based upon the size and occurrence of the plagioclase, clinopyroxene and olivine phenocrysts which is easily applied to field mapping (MacGregor, 1928). Names are based upon type localities in central and southern Scotland

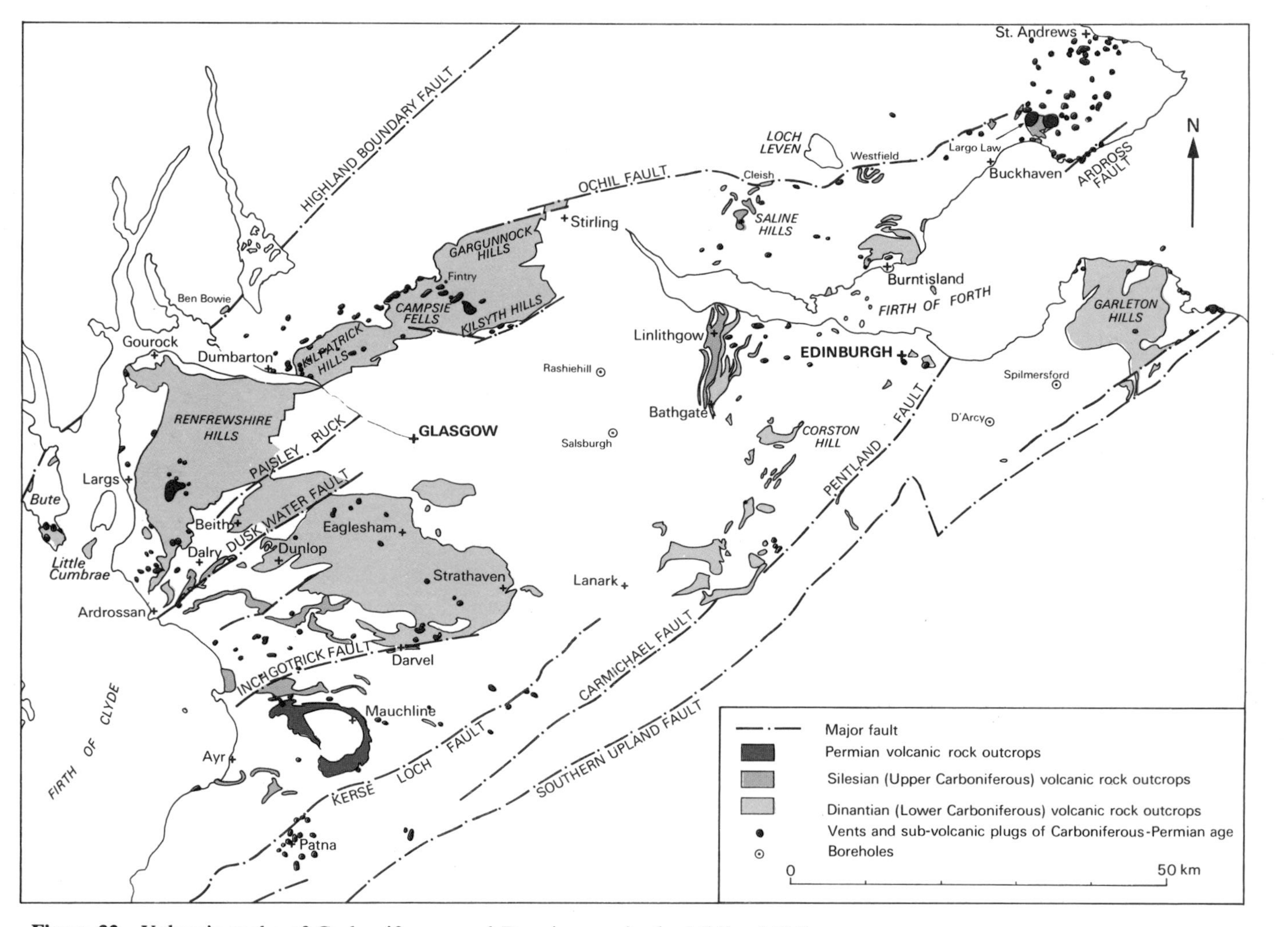

Figure 32 Volcanic rocks of Carboniferous and Permian age in the Midland Valley

and the classification is still in active use. Although some of MacGregor's petrographic categories include a range of petrochemical compositions, a general correlation is possible with currently accepted terminology (Table 5; Macdonald, 1975). Notable departures from a true 'basalt' composition are the very mafic flows (Craiglockhart type) which are ankaramitic, the feldspar-phyric flows (Jedburgh and Markle types) which are often hawaiites and some Hillhouse types which are basanites.

The mugearites are typically fine-grained and aphyric with a platy parting parallel to a planar flow-alignment of plagioclase crystals. Titanomagnetite and apatite are more abundant in rocks of intermediate composition and hornblende phenocrysts are sometimes present in the trachybasalts and trachyandesites. Trachytes and rhyolites may contain phenocrysts of sanidine, quartz and rarely biotite, hornblende or augite.

Table 5 Nomenclature of basic igneous rocks of Carboniferous and Permian age in the Midland Valley

Basalt type of MacGregor (1928)	Phenocrysts		Chemical classification of Macdonald (1975)	Type locality
	abundant	sometimes present in lesser amounts		
Macroporphyritic (phenocrysts >2 mm)				
Markle	plag	±ol, Fe-oxide	plag ±ol ±Fe-oxides-phyric basalts, basaltic hawaiites or hawaiites	Markle Quarry, East Lothian (flow)
Dunsapie	plag + ol + cpx	±Fe-oxide	ol-cpx-plag-Fe-oxides-phyric basaltic hawaiites, or ol-cpx-plag-phyric basalts	Dunsapie Hill, Edinburgh (vent intrusion)
Craig-lockhart	ol + cpx		Ankaramite	Craiglockhart Hill, Edinburgh (flow)
Microporphyritic (phenocrysts <2 mm)				
Jedburgh	plag	±ol, Fe-oxide	plag±ol ±Fe-oxides-phyric (basaltic) hawaiites, (occasionally basalt)	Little Caldon, Stirlingshire (plug). Also in Jedburgh area
Dalmeny	ol	±cpx, plag	ol±cpx-phyric basalt	Dalmeny Church, West Lothian (flow)
Hillhouse	ol + cpx		ol-cpx-phyric basalt (rarely basanite)	Hillhouse Quarry, West Lothian (sill)

plag = plagioclase ol = olivine cpx = clinopyroxene

The range of lava types present in any one area varies considerably but Macdonald (1975) defines three main volcanic associations. The first includes a full range of compositions from ankaramite to trachyte such as occurs in East Lothian and most of the southern Clyde Plateau. The second is characterised by feldspar-phyric hawaiites and mugearites with local trachytes as in the north-eastern part of the Clyde Plateau. The third has a restricted range of ankaramites and olivine- or olivine-pyroxene-microporphyritic basalts such as occurs in most of the Namurian and younger suites.

The basaltic lavas can be remarkably fresh, although olivine is usually replaced by red-brown pseudomorphs. Less fresh material shows varying degrees of albitisation, chloritisation, oxidation, hydration and replacement by carbonate. Albitisation in particular can lead to considerable difficulties in petrographic and petrochemical classification. A high proportion of many flows consists of zones of amygdaloidal material, autobrecciated and/or hydrothermally altered rubble and slaggy, vesicular flow tops. Typical amygdale and vein assemblages involve combinations of chlorite, hematite, calcite, quartz and chalcedony. Zones of intense, possibly penecontemporaneous, hydrothermal alteration commonly contain a range of zeolites and related minerals (e.g. stilbite, heulandite, analcime, prehnite, apophyllite, pectolite) and native copper is known from several localities (e.g. Boylestone Quarry, Barrhead). The more differentiated rocks are usually heavily altered throughout and silicification is common in the rhyolites.

Dinantian volcanic activity

The earliest widespread activity in the Midland Valley is of early Viséan age. Thick lava sequences are developed within the lower Calciferous Sandstone Measures of East Lothian and Edinburgh City and thinner developments may be traced south-westwards towards Carstairs. The thickest and most extensive development is in the west where the 'Clyde Plateau Lavas' form a continuous outcrop to the north, west and south of Glasgow (Figure 32). East of Glasgow, the lavas are known to be present beneath younger sediments and thinner sequences occur further west on the islands of Bute and Little Cumbrae. In the Bathgate Hills of West Lothian and in the Burntisland area of Fife, the main volcanic accumulations commenced in the Upper Viséan and continued into the Namurian.

Many of the Viséan volcanic centres were initiated in a shallow marine or coastal plain environment and sequences often commence with bedded pyroclastic material and/or volcaniclastic sediments. However, in most areas a rapid accumulation of volcanic deposits created subaerial lava plateaux, which then remained above sea level throughout the volcanic episode, so that intercalated sediments are rare or non-existent. Tropical weathering between eruptions in some areas has produced reddened flow tops or persistent red-brown lateritic boles. Individual flows vary in thickness between 5 and 30 m and are usually of aa type, although rare pahoehoe features have been reported.

Eruptions are thought to have been from relatively small and short-lived central volcanoes, the remains of which are preserved in some areas as vents and plugs, usually of less than 500 m diameter, surrounded by proximal lava flows and pyroclastic features. Away from such centres, pyroclastic rocks are rare and lava plateaux occur which may have been fed, at least in part, from fissure eruptions. In some areas vents, plugs and local dyke swarms are grouped together in linear zones with a predominant NE–SW trend, which may reflect deep-seated fractures along Caledonian structures in the underlying basement. Several such fractures exerted a major structural control during the Viséan and marked changes in thickness of both sedimentary and volcanic successions occur across NE–SW faults such as the Dusk Water Fault. Other, less well defined volcanic lineaments have been postulated on a NW–SE trend in the western Midland Valley where such trends become more important during later igneous activity.

In East Lothian the volcanic sequence of the Garleton Hills is up to 520 m thick. Basal basaltic tuffs are interbedded with thin, lagoonal sediments (200 m) and are succeeded by ankaramite, basalt, hawaiite and mugearite flows (160 m) and an upper group of thick trachyte flows and tuffs (160 m). Similar but thinner sequences are observed in the Spilmersford Borehole (250 m) and in the D'Arcy Borehole (75 m) (Figure 32). The basic lavas are mainly of Craiglockhart, Dunsapie and Markle type and intermediate types include kulaites (trachybasalts with hornblende phenocrysts and analcime, possibly as pseudomorphs after leucite). The more salic rocks are mostly quartz-trachytes but locally include quartz-banakites (quartz-bearing trachyandesites with phenocrysts of green sodic augite), tuffs and rare welded tuffs. Agglomerate-filled vents are particularly abundant and well exposed in coastal sections where the basal pyroclastic unit reaches its thickest development. Some contain basaltic minor intrusions and larger sills, laccoliths and plugs of trachyte, phonolitic trachyte and phonolite form prominent landmarks such as the Bass Rock, North Berwick Law and Traprain Law. Such intrusions are believed to belong to the same eruptive phase as the lavas and this is supported by K-Ar dating.

In Midlothian the sequence of lava flows and tuffs which form Arthur's Seat and Calton Hill in the centre of Edinburgh occur at a similar horizon to the East Lothian volcanic rocks (Figure 33). The 400 to 500 m-thick lava and pyroclastic sequence at Arthur's Seat comprises some 13 flows, with several well defined tuff bands. Lower flows are basalts and ankaramites of Dunsapie and Craiglockhart type with Markle and Jedburgh types and mugearites above. At Calton Hill, a similar sequence is only 200 m thick. The area was a major centre of activity with three large, agglomerate-filled vents, two on Arthur's Seat (the Lion's Head and Lion's Haunch Vents) and one on Salisbury Craigs (the Western Vent). Blocks of lava within the vents have been matched with specific local flows and comparable rock types also occur in numerous plugs, sills and dyke-like masses. The basalt plug of Edinburgh Castle Rock, 3 km to the west-north-west, may be part of the same activity.

A 60 m-thick sequence of tuffs and basalt at Craiglockhart Hill, 6 km to the south-west of Arthur's Seat, rests upon red sandstones similar to those which occur at various levels in the Cementstone Group. The volcanic rocks have traditionally been placed near the base of the local Carboniferous succession but could equally be near-contemporaneous with the Arthur's Seat activity. Further to the south-west thin sequences of basalt, mugearite and tuff occur between Corston Hill and Crosswood Reservoir. These outcrops are almost continuous with wider outcrops resting upon rocks of Devono–Carboniferous facies or directly upon Lower Devonian rocks, in poorly exposed ground around the south-western end of the Pentland Hills anticline and around

Carstairs. Thin basaltic lava flows occur at a higher stratigraphical level in the Midlothian syncline at Carlops.

In the western Midland Valley the Clyde Plateau volcanic rocks occur in several fault-bounded blocks, each with its own characteristics and between which only tentative correlations may be made. Limited stratigraphical evidence suggests that lava plateaux, up to 900 m thick in places, accumulated within a relatively short space of time (Figure 33).

The volcanic sequences overlie rocks ranging from the Upper Devonian to high in the Cementstone Group of the basal Carboniferous and hence it seems likely that a period of uplift and erosion preceded the initial volcanic outpourings. Following the termination of eruptive activity, continuing subsidence allowed upper Dinantian sediments to encroach upon the newly-created volcanic terrains and volcaniclastic detritus eroded from such areas commonly overlies the lavas.

Volcanic sequences in the north-east from the Touch Hills to the Campsie Hills are 400 to 500 m thick and are composed almost entirely of feldspar-phyric hawaiites (Jedburgh and Markle type) with subordinate mugearites and trachybasalts at higher stratigraphic levels in some areas. To the south-west a similar succession may be recognised in the Kilpatrick Hills, with additional, more-basic lower and upper units of basalts (Dalmeny and Dunsapie type) and ankaramites (Craiglockhart type) in a total thickness of about 400 m.

Evidence for the existence of eruptive centres is abundant throughout the northern Clyde Plateau. Volcanic rocks and underlying sediments are pierced by numerous agglomerate-filled vents, with or without vent intrusions, and by small cylindrical plugs of basalt which form prominent landmarks (Plates 6.2 and 9.2). Close to the vents, lava flows are often subordinate to bedded pyroclastic rocks, some of which may be the degraded remains of lines of ash cones. Many of the vents lie on prominent NE–SW lines within a 2 to 3 km wide zone extending 27 km from Fintry to Dumbarton (including Dumbarton Rock, Dumgoyne and Dunmore). Others occur along an 8 km length of the Campsie Fault to the south-east of the Kilsyth Hills. Less well defined lines of vents trend NW–SE, one through Bowling, east of Dumbarton, and one in the Campsie Hills from Dunmore to Meikle Bin. The Meikle Bin vent is thought to lie on the margin of a caldera within which the rocks are brecciated and heavily metasomatised and an underlying major basic intrusion is indicated by geophysical evidence. The presence of acid intrusions and trachytic agglomerate within the vent indicates that more-salic lavas were erupted, although none are now preserved. A small intrusion of phonolitic trachyte occurs at Fintry.

To the south-west of Dumbarton, vents and plugs are rare in the northern part of the Renfrewshire Hills, where the general form is of a widespread lava plateau, probably fed from fissure eruptions. However, where the lower part of the lava pile is exposed in deep valleys, numerous dykes with a predominant NE–SW trend occur, particularly along the projected continuation of the Dumbarton–Fintry line. Comparable dyke swarms, vents and plugs are also seen along a further projection of the line, cutting Upper Devonian and basal Carboniferous sediments on Great Cumbrae, and the lava succession of South Bute.

The succession in the Renfrewshire Hills may be as much as 800 m thick, consisting largely of an alternating sequence of Markle-type hawaiites and

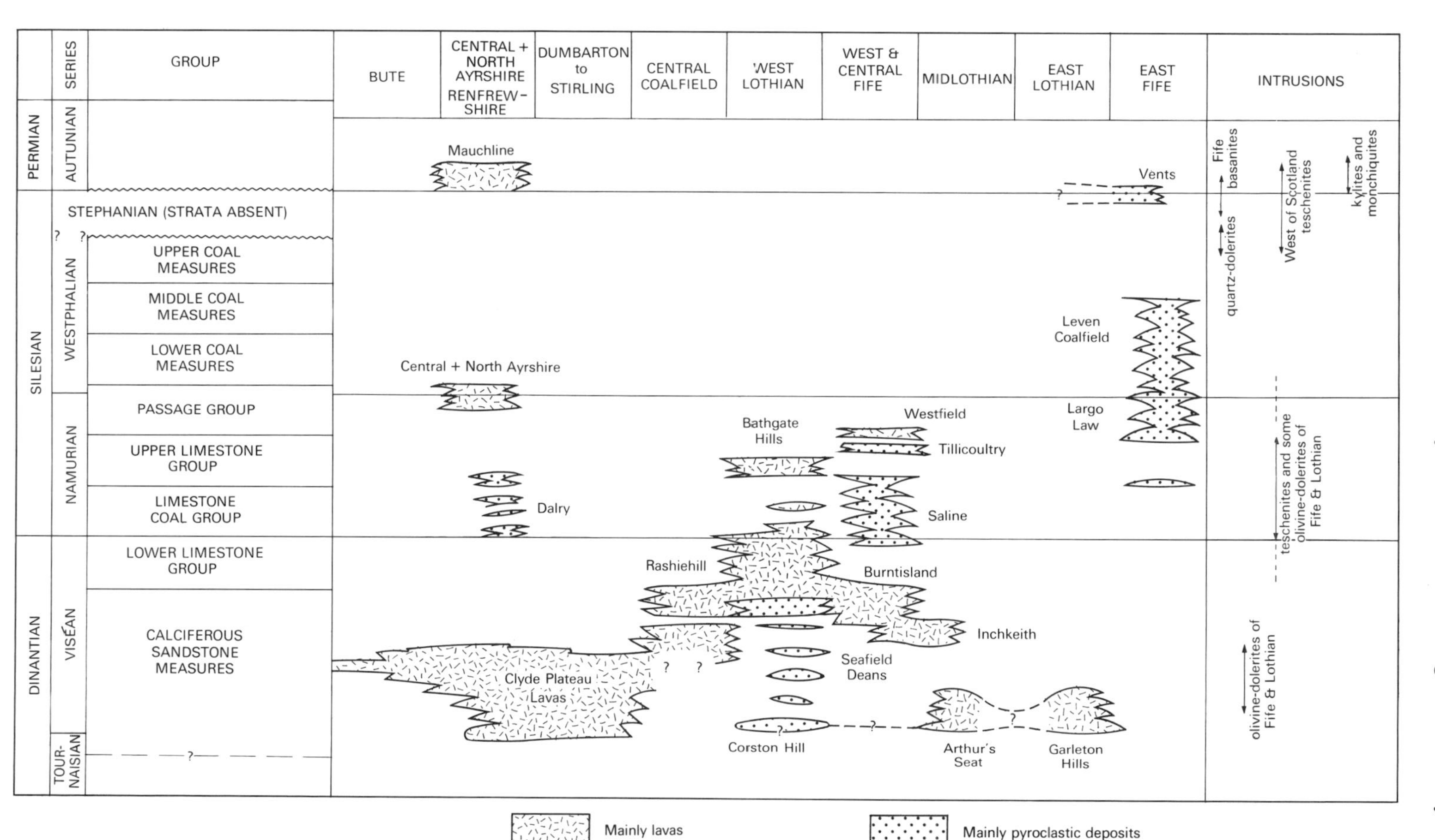

Figure 33 Diagrammatic representation of geographical and stratigraphical distribution of Carboniferous and Permian igneous rocks of the Midland Valley

mugearites. Basaltic tuffs, ankaramites and microporphyritic basalts form a basal unit and Dunsapie and Dalmeny type basalts predominate in a well-defined, upper series. In the middle of the main series a thick pile of trachyte and rhyolite lavas, resting upon trachytic agglomerate and cut by trachytic plugs form the Misty Law Centre. Large trachytic vents occur at Irish Law and Knockside Hill and trachytic dykes are common in the underlying dyke swarm.

The Renfrewshire Hills volcanic succession thins southwards towards Ardrossan where numerous sills, small bosses and agglomerate-filled vents also occur. Some of these are contemporaneous with the lavas, others may be of Namurian age and one group is believed to be Permian. The Heads of Ayr vent some 25 km further south, cuts bedded agglomerates at the top of the Cementstone Group, and may be an isolated occurrence of Dinantian activity, although vent intrusions have more in common with the Permian vents of the area. Westwards, 180 m of Markle and Jedburgh type hawaiites with subordinate mugearites occur on Little Cumbrae and a sequence of comparable thickness on south Bute ranges in composition from basalt (Dunsapie type) to trachyte.

The southern part of the Clyde Plateau volcanic outcrop extends east-south-east from the Renfrewshire Hills to Strathaven and is divided into several blocks by major NE–SW faults such as the Dusk Water Fault. Geophysical investigations have suggested that the volcanic succession is thickest, possibly up to 900 m, along the WNW–ESE axis of this outcrop, but that marked changes in thickness occur across the NE–SW faults (Hall, 1974). The lavas thin abruptly in the south-east at the Inchgotrick Fault, south of which only a few thin flows are known from the Sorn area. A wide range of basic and intermediate lavas are represented, but microporphyritic basalts of Dalmeny type are particularly common. The Dunlop–Eaglesham–Darvel Hills are characterised by the presence of a wide range of salic lavas including hornblende-trachyandesites, trachytes, quartz-trachytes and rhyolites, together with associated bedded pyroclastic material (Plate 6.1). Over 20 plugs and vents, mostly of trachyte and phonolite are broadly distributed along an ESE axis between Irish Law and Loudoun Hill.

The Clyde Plateau volcanic rocks are assumed to be continuous beneath the Central Coalfield Syncline in the Glasgow area and probably extend eastwards as far as the Rashiehill Borehole at Slamannan and the Salsburgh No 1A oil well. Further east they are thought to thin rapidly and are replaced by the thick sedimentary succession of the West Lothian oil-shale basin. The original extent of the Clyde Plateau may therefore have been some 3000 km^2.

In West Lothian most of the volcanic activity is younger than the Clyde Plateau and Arthur's Seat sequences. The Crosswood Ash (100 m thick), the Seafield–Deans Ash (250 m thick) and numerous thin, widespread tuffs occur in the middle of the Calciferous Sandstone Measures. Bedded tuffs and agglomerates become more persistent in the Bathgate Hills in the upper Calciferous Sandstone Measures. The tuffs are succeeded by basaltic and basanitic, microporphyritic lavas of Dalmeny and Hillhouse type, with subordinate pyroclastic rocks, which constitute most of the Lower Limestone Group. Intercalated sedimentary sequences show that, in contrast to the Clyde Plateau, the balance between eruption, erosion and deposition was periodically reversed. Many of the volcanic episodes formed ephemeral islands in a shallow sea and in some cases fringing limestone reefs have been identified. Numerous necks and plugs which cut the West Lothian oil-shale field suggest a former eastwards extension of the volcanic field, which may also be traced westwards, beneath the Central Coalfield, at least as far as the Rashiehill Borehole. The volcanic activity continued into the Namurian resulting in a total of up to 500 m of volcanic rocks.

In Fife a volcanic pile, similar in petrography and overall character to that of the Bathgate Hills, constitutes much of the upper Calciferous Sandstone Measures in the closure of the Burntisland Anticline. Thin, sedimentary intercalations are common in a sequence of Dalmeny and Hillhouse type basalt lavas with subordinate tuffs. The sequence is 450 m thick at Burntisland and Seafield Colliery and extends eastwards to Inchkeith Island where it is at least 150 m thick. Several necks and plugs, consisting of similar petrographic types to the lavas, are assumed to be contemporaneous, the most conspicuous being the Binn of Burntisland. Small outcrops of volcanic rocks, including some rhyolites, in the Cleish Hills of west Fife are of Dinantian age.

Silesian volcanic activity

Volcanism continued almost continuously throughout the Silesian, but at a less productive level than in the Dinantian. Activity was concentrated in relatively short-lived, local centres where, in most cases, phreatic explosive eruptions from a multitude of complex vents produced an abundance of bedded pyroclastic deposits within shallow-water sedimentary sequences.

Surface lava flows are predominant only in Ayrshire, but large volumes of magma solidified at depth in most areas as sill-complexes. A restricted compositional range of basalts and basaltic hawaiites, with some basanites, is in marked contrast to the wide range of differentiates present in most Dinantian sequences. The more explosive nature of the volcanism may be caused by changes in magma composition and tectonics, but could also be related to the increasing thicknesses of geotechnically weak sediments in the rapidly-developing Silesian basins. In general, such sequences would be of too low density to support a column of magma, which would spread laterally in the form of sills. Magmas which did have sufficient energy to rise to shallower levels would react with wet sediments to produce violent phreatic eruptions.

The Bathgate Hills sequence of basalt lava flows with subordinate tuffs is continuous from the Dinantian, into the Limestone Coal Group, which is almost completely replaced by lavas around Linlithgow. In the Upper Limestone Group, between 40 and 180 m of lavas occur, mainly between the Index and Orchard limestones, but they extend almost up to the Calmy Limestone in their thickest development. In the northern part of the Central Coalfield, 15 m of lavas and tuffs have been proved in boreholes above the Calmy Limestone and tuff bands occur at several horizons up to the lower Passage Group.

In Fife, evidence of Silesian volcanicity in the form of interbedded tuffs, tuffaceous siltstones and small diatremes is widespread. The tuffs consist of a mixture of basaltic and comminuted sedimentary debris and are often well-graded, indicative of ash-fall into shallow water. Diagenetic alteration produces kaolinised tuffs or tonsteins which are especially common in or near to coal seams. Volcanic activity is particularly concentrated in three areas. In the Saline Hills of west Fife thick tuffs and rare, thin basalt flows in the Limestone Coal Group and parts of the Upper Limestone Group are cut by several necks and plugs. In central Fife, five flows of basaltic pillow lava with associated tuffs and hyaloclastites are interbedded with the top of the Upper Limestone Group and basal Passage Group around Westfield opencast coal pit, near Kinglassie. In east Fife, bedded tuffs and a few lavas are associated with the Upper Limestone Group and Passage Group around the large complex vents of Largo Law and Rires. Over 100 smaller necks, many with plugs of Hillhouse-type olivine-basalt, cut almost the full local age range of Dinantian and Namurian strata. It is therefore inferred that at least some of these necks are contemporaneous with the Namurian bedded tuffs. However, a few small necks cut Lower and Middle Coal Measures strata and one, the Viewforth neck, contains sediments with Westphalian A spores. Some necks are thus of Westphalian age or younger and may be contemporaneous with a sequence of over 170 m of tuff, agglomerate and rare basalt encountered in offshore boreholes near Leven which range from topmost Passage Group to Middle Coal Measures (Westphalian B).

Some of the vent agglomerates, including those at Largo Law and Rires, are cut by plugs and minor intrusions of basanite, olivine-analcimite or olivine-nephelinite (formerly termed monchiquites) and hence resemble the Permian age vents of Ayrshire. Indirect evidence suggests that these vents post-date the late-Westphalian/early-Stephanian quartz-dolerite suite. Whole-rock K-Ar dates obtained from the freshest basanites and monchiquites fall within a narrow range of 290 to 280 Ma suggesting that the vent intrusions are of late-Stephanian or early-Permian age.

Monchiquitic and basanitic rock-types also occur in vent intrusions, stocks and sills of the south side of the Firth of Forth, where they cut Viséan lavas and overlying sediments. In the absence of any evidence to the contrary it seems reasonable to suppose that these intrusions are a continuation of the east Fife late-Stephanian/early-Permian province, from which they are separated by only 15 km. Notable examples of intrusions occur at Kidlaw, Limplum and around North Berwick. It is probable that some of the numerous agglomerate-filled necks in the area belong to the same phase of activity.

The Fife vents have been the subject of much detailed work on their internal structure, mode of emplacement and relationships to surrounding bedded tuffs and contemporaneous sedimentation. Most appear to be funnel-shaped tuff-pipes, giving rise to Surtseyan eruptions and modified by post-eruptive subsidence and inward collapse of the area surrounding the initial pipe. Structural control of vent sites is well illustrated by the NE–SW Ardross Fault, along which ten or more vents are sited in a distance of 4 km from Elie to St Monance. A number of vents contain megacrysts and/or rock clasts of deep-seated igneous material or metamorphic basement (pp.7–8, 112).

In Ayrshire, tuffs and agglomerates occur throughout the Limestone Coal Group in the area around Dalry, replacing locally the Dalry Blackband Ironstone and large parts of the coal-bearing succession. Less extensive tuffs occur in the Upper Limestone Group and all may have been derived from necks which cut older strata to the west.

In the Passage Group, widespread volcanism produced a series of basalt lavas, which underlie much of the Ayrshire Coalfield Basin (Figure 32). The sequence of lavas with minor intercalated sediments thickens from 10 m on the basin margins to 160 m in the Barassie inlier, near Troon. It occupies a stratigraphical position at the junction of the Passage Group and Lower Coal

Measures and hence is perhaps partly Westphalian in age. The lavas are almost exclusively microporphyritic olivine-basalts of Dalmeny type but have been shown to have some tholeiitic to transitional chemical tendencies.

These lavas were erupted on to a flat, subsiding land surface, which was frequently submerged, allowing shallow-water lagoonal sediments and coals to be deposited between flows. Other flows show many indications of long periods of weathering between eruptions under hot and humid climatic conditions which led to the formation of ferruginous or aluminous clays, either as in situ lateritic residual deposits, or by a combination of sedimentary reworking and chemical precipitation from solutions charged with aluminium hydroxide. Following the last eruptions, a prolonged period of emergence and deep weathering gave rise to the Ayrshire Bauxitic Clay which reaches thicknesses of up to 9 m at the top of the lava pile in the northern outcrops.

Permian volcanic activity

Interbedded volcanic rocks of Permian age occur only in Ayrshire, where they form an annular outcrop around the overlying aeolian sandstones of the Mauchline Basin. Contemporaneous necks and sub-volcanic intrusions are abundant within and around the Mauchline lavas, and the late necks of east Fife and possibly East Lothian have late-Stephanian to Lower Permian radiometric ages. Sills of various alkali dolerite rock-types are thought to be near contemporaneous with the Lower Permian lavas.

The Mauchline volcanic sequence, which increases in thickness eastwards from 100 to 238 m, rests unconformably but with no marked discordance upon Upper Coal Measures. Sediments resembling the overlying Mauchline Sandstones occur locally at the base of the sequence, as intercalations between flows and as infillings in slaggy flow tops. Lava flows are predominantly olivine-basalts of Dalmeny type, but some strongly silica-undersaturated basic types are characteristically present (nepheline-basanite, analcime-basanite, olivine-nephelinite) and analyses include some hypersthene-normative (transitional) basalts. Agglomerates and tuffs constitute a large part of the succession, becoming more abundant in the thicker, eastern parts.

Over 60 volcanic necks are known in Ayrshire, mostly within a 20 km radius of the Mauchline Basin, but also extending to West Kilbride in the north, Muirkirk in the east and Dalmellington in the south. Numerous lines of evidence suggest that these necks are contemporaneous with the Mauchline lavas and hence delimit the former extent of the volcanic field. Necks are known to cut the Coal Measures, post-Coal Measures alkali dolerite sills and the Mauchline lavas but not the Mauchline Sandstones. Those which cut older strata do so in areas where older volcanicity is not recorded. Many vents contain wind-rounded sand grains and some include large subsided blocks of Mauchline sandstone.

Vent agglomerates usually consist of a mixture of sedimentary and igneous material. Blocks and lapilli of Mauchline lava types are common and alkali dolerite, derived from sills, occurs locally. Plugs and other vent intrusions are predominantly of highly-undersaturated olivine-analcimite or monchiquite but alkali dolerite and camptonite are also known. Thin dykes and sills of monchiquite commonly occur in the vicinity of necks and monchiquite dykes are common in the Irvine valley and the Patna area. Many of the intrusions and agglomerates contain xenolithic megacrysts and nodules of carbonated peridotite.

Inclusions in volcanic and sub-volcanic rocks

A wide variety of xenolithic igneous inclusions and megacrysts are found in many intrusions, pyroclastic deposits and more rarely lavas, particularly in east Fife, East Lothian and Ayrshire. The nodule suites are regarded as broadly contemporaneous with their host rocks and this has been supported by radiometric age determinations in east Fife. Particularly well-documented suites have been extracted from vents at Partan Craig, East Lothian and Elie Ness, east Fife.

In addition to fragments of metamorphic continental crust (Chapter 2) and contemporaneous lavas, often of a basanitic or monchiquitic nature, inclusions of ultramafic and ultrabasic material are relatively abundant. Magnesian peridotites, usually showing a metamorphic texture, may represent upper mantle wallrock depleted by partial melting episodes. Undeformed rocks may represent slightly younger intrusions within the mantle and consist of a wide spectrum of clinopyroxene-bearing ultramafites, orthopyroxenites and garnet-spinel-peridotites. In some areas (e.g. East Lothian) the pyroxenites contain potassic hydrous minerals such as amphibole (kaersutite) and biotite, suggestive of potash metasomatism within the mantle. Ultrabasic and basic rocks, often with a cumulus texture, consist of combinations of plagioclase, olivine, augite, kaersutite and biotite. Such minerals also occur as disaggregated megacrysts along with titaniferous garnet (known locally as 'Elie ruby'), zircon, corundum and apatite in assemblages suggestive of a high pressure origin.

Higher level, crustal or sub-crustal magma chambers are a more likely source for xenoliths of layered gabbro and picrite or more rarely syenite, and for megacrysts of anorthoclase.

Alkali dolerite and related sills

Large sills and sill-complexes of basic alkaline rocks occur within the Carboniferous basins of the Lothians and Fife, in the western part of the Central Coalfield basin around Paisley and Glasgow, and in the Ayrshire Coalfield basin (Figure 34). There is no associated dyke swarm. Most are of Silesian to Permian age and may occur at deeper levels in basins of thick sediments, where near-contemporaneous surface eruptions are of a predominantly explosive, pyroclastic nature. In coalfield areas they are known to intrude along coal seams. In such circumstances, the coal may be either totally replaced, 'burnt' or coked, or converted to a higher-grade anthracitic coal and the dolerite is altered to 'white trap'. The sills have been quarried extensively for aggregate throughout the Midland Valley.

Petrography of the alkaline sills

Basic alkaline rocks of the Midland Valley were the subject of numerous early works on the origin and differentiation of basaltic magmas, on account of the wide variety of rock-types often present within a single composite or differentiated body such as the Lugar and Saltcoats sills of Ayrshire and the Braefoot Outer Sill of Fife. Almost all the sills are of olivine-bearing doleritic

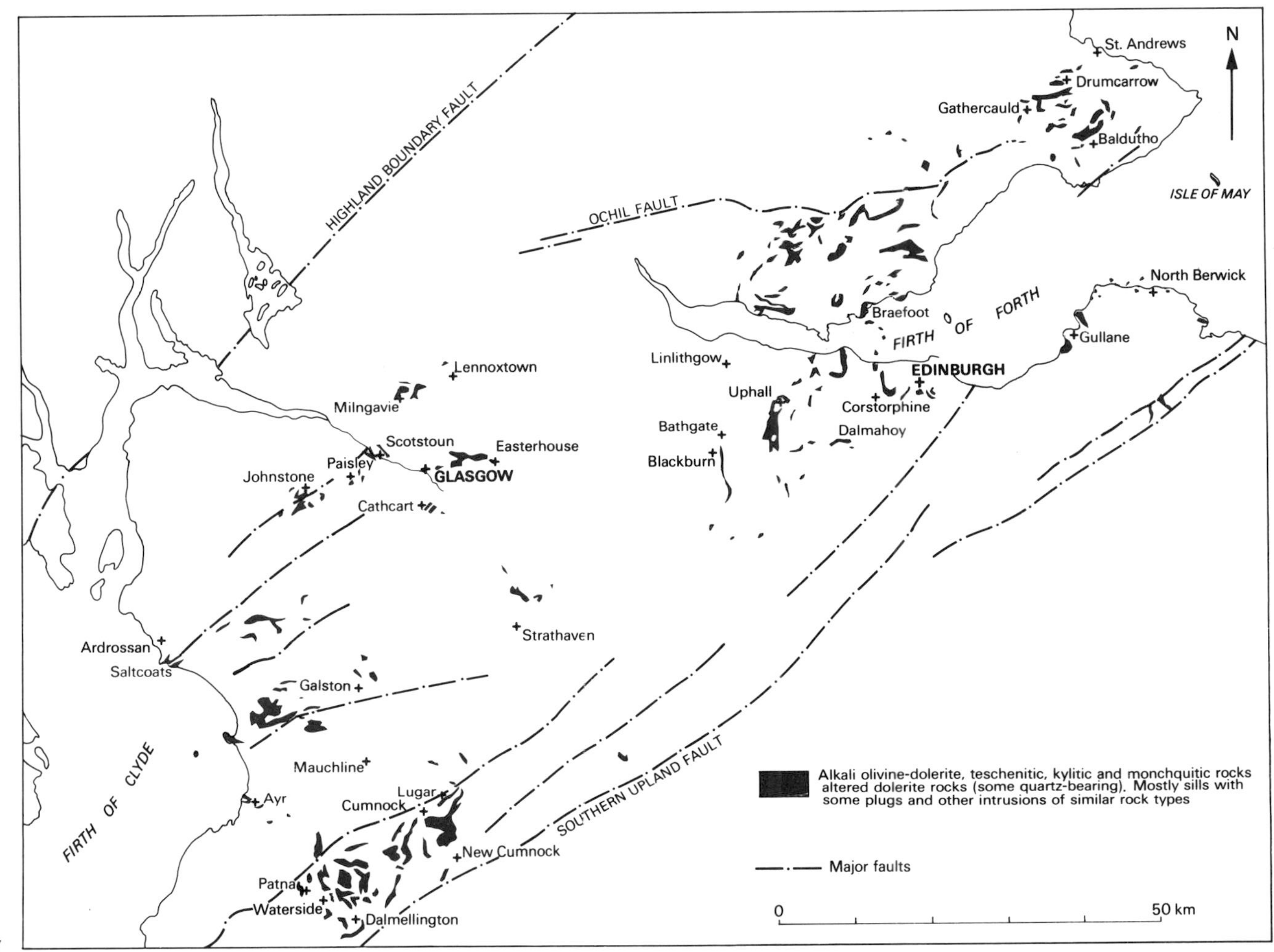

Figure 34
Alkali dolerite sills and associated intrusions of Carboniferous and Permian age in the Midland Valley

rock types which have been divided for descriptive purposes into five groups. Within these groups many individual names have been utilised to describe the more-distinctive rock-types. Composite or differentiated intrusions rarely contain rock-types from two or more groups and many include basic or ultrabasic layers in their lower parts.

1 Kylitic types are characterised by the presence of abundant olivine (10–40%), usually with nepheline and euhedral, zoned, purplish augite. They include essexite, theralitic essexite, theralite, olivine-rich theralite (= 'kylite'), basanite and picrite.

2 Teschenitic types have less olivine (5–15%) than the kylitic types and all are analcime-bearing. They include teschenite, hornblende-teschenite, camptonite, bekinkinite, picrite and peridotite. Most of the rocks described as 'teschenite' in the Midland Valley are strictly 'olivine-bearing teschenites'.

3 Alkali olivine-dolerites and basalts are mildly-undersaturated but with no visible nepheline and little analcime. Textures may be ophitic, sub-ophitic, intergranular or microporphyritic, the latter commonly resembling the Dalmeny lava-types.

4 Monchiquitic types are fine-grained, feldspar-free rocks consisting of phenocrysts of olivine and augite in a mesostasis of glass, analcime or nepheline. Most lack the amphibole, characteristic of true monchiquites and thus are more correctly termed olivine-analcimites or olivine-nephelinites. Similar rocks containing some plagioclase are analcime-basanites, nepheline-basanites and leucite-basanites. Many are xenolithic (p.112).

5 Altered doleritic types include chloritised and serpentinised olivine-dolerites, some with residual analcime and some with juvenile quartz. Others are altered olivine-free dolerites with traces of quartz.

A characteristic of both alkali dolerite and tholeiitic quartz-dolerite intrusions throughout the Midland Valley are the zones of 'white trap' in which the normal rock is transformed into a pale white, cream or yellowish brown alteration product. The primary doleritic texture is usually preserved, but the constituent minerals are pseudomorphed by kaolinite, chlorite, leucoxene, amorphous silica and carbonate. Most 'white trap' is associated with fault planes which can contain vein mineralisation (Chapter 16). 'White trap', commonly containing solid or viscous hydrocarbons on joint surfaces, is particularly widespread in dolerites that are associated with carbonaceous shales, coals or oil-shales. It has been suggested that the alteration is caused by volatiles released during the distillation of such rocks by heat from the intrusions.

Dinantian to early Westphalian sills of the Lothians and Fife

Major sills up to 120 m thick, in the eastern Midland Valley are mostly of olivine-dolerite or teschenitic type, although monchiquitic types occur as smaller bodies in the later Stephanian/early-Permian vents of east Fife and East Lothian.

The sills cut mainly Calciferous Sandstone Measures in the Lothians and extend into the Upper Limestone Group in Fife, but are absent from the Passage Group and Coal Measures. It has therefore been suggested that they are of late-Viséan to Namurian age, contemporaneous with the volcanicity of the area. Quartz-dolerite dykes of the early-Stephanian regional swarm cut teschenite sills on Inchcolm Island and near Linlithgow and hence provide an

1 Quartz-dolerite dyke of late-Carboniferous swarm showing lateral shift in alignment, Campsie Linn, River Tay. (D 3310)

2 'Rock and Spindle' sea stack near St Andrews. Resistant mass of xenolithic, columnar basanite and basanitic breccia in softer tuffs of the late Carboniferous or early Permian Kinkell Ness volcanic vent. (D 1764)

Plate 11

upper age limit. Some of the olivine-dolerite sills bear a strong petrographic and geochemical resemblance to neighbouring extrusive rocks and may be of Viséan age. The teschenites, however, form a separate distinctive suite and appear to be later, although their fine-grained margins often consist of olivine-basalt or basanite, akin to the Dalmeny type lavas of the Bathgate Hills and Burntisland. Veins of sediment within several teschenite sills were originally taken to indicate intrusion into unconsolidated sediments during the Viséan, but some veins have been reinterpreted as intrusive tuffisites, possibly related to later volcanic activity. Recent K-Ar whole-rock dates suggest minimum ages within the range 313 to 296 Ma (i.e. Westphalian) and hence imply affinities with the Namurian to early-Westphalian volcanicity.

Teschenitic sills are widespread in the Lothians, including the sills of Gullane, Salisbury Craigs, Craigie and Blackburn. Altered dolerites form the Corstorphine Sill, Edinburgh and the Stankards Sill, Uphall, both of which contain thick bands of picrite. Rocks from the Lochend Sill, Edinburgh and Craigleith sill, North Berwick have been described as 'essexites'. An olivine-bearing doleritic sill with some tholeiitic affinites in the Dalmahoy and Kaimes area of West Lothian has previously been regarded as an atypical member of the Stephanian quartz-dolerite suite. However, a K-Ar date suggests a minimum age of 320 Ma and supports an earlier interpretation that the sill is contemporaneous with the Dinantian or Namurian volcanic activity.

In Fife olivine-dolerite and teschenite sills are abundant, forming sill-complexes up to 115 m thick which are well known from coal workings and boreholes, as well as from extensive surface outcrops. The distribution of the various types shows a rude zonal pattern in east Fife which is independent of geological structure. Examples of ophitic, non-ophitic and microporphyritic olivine-dolerites occur in major sills at Gathercauld, Drumcarrow and Baldutho. Well-known examples of teschenitic types occur on the Isle of May, on Inchcolm Island and at Braefoot Point. The latter has a layered structure attributable to gravitative sinking of olivine, and consists of picroteschenite, teschenite, dolerite and dolerite-pegmatite within chilled margins of basalt.

Westphalian to Permian sills of the western Midland Valley

Representatives of almost all the rock-types listed on p.114 occur in extensive sills and sill-complexes in the Ayrshire Coalfield basin and teschenite sills are abundant in the Glasgow–Paisley area. Representatives of all types cut Coal Measures strata and most sills are therefore of late-Westphalian age or younger age.

In the Ayrshire Coalfield the kylitic and monchiquitic sills and some teschenites are younger than all major faults of the area. They have strong petrographic and geochemical affinities with the Lower Permian Mauchline lavas, are cut by necks and dykes associated with the lavas, and occur as blocks in vent agglomerates. They are therefore assumed to be slightly older than or broadly contemporaneous with the Permian lavas. Palaeomagnetic measurements support a Permian age and K-Ar mineral dates indicate minimum ages in the range 285 to 276 Ma. The majority of teschenitic sills (including the Lugar Sill), teschenitic olivine-dolerites and a variety of altered doleritic types also cut the Coal Measures and most major faults, but are affected by a set of NW–SE-trending faults. These sills are therefore assumed to be slightly older and separated minerals have yielded K-Ar minimum ages in the range 297 to 286 Ma. Most are probably of late-Westphalian or early-Stephanian age, but it is possible that some (e.g. the Craigie Sill, south of Kilmarnock) are associated with the Passage Group lavas.

Kylitic sills are well developed in the Kyle district of Ayrshire, north of Dalmellington. The Benbeoch sill includes the type 'kylite' (an olivine-rich theralitic essexite) and contains picritic layers. Other picritic sills occur locally and felsic, alkaline segregation veins are recorded from Kilmein Hill. Monchiquitic sills are never more than 2 m thick and are closely associated with volcanic necks of the Mauchline lavas (p.111). Notable examples occur in the Waterside area south of Patna.

Teschenite sills are numerous in the area around Cumnock and also between Ardrossan and Galston. Other sills of less-alkaline teschenitic olivine-dolerite occur mainly in the area between Patna, Dalmellington and Cumnock. The Lugar sill is thought to have formed from an initial intrusion of picroteschenite magma followed shortly afterwards by one or more pulses of olivine-theralite or peridotite and later veins of lugarite (amphibole-rich ijolite). There has been much debate as to whether differentiation took place before or after intrusion. Other composite intrusions which include both kylitic and teschenitic rock types occur in the Patna area.

Highly-altered olivine-dolerites, some with secondary quartz, form extensive outcrops near Dalmellington and Cumnock. They include the Craigens–Avisyard composite sill-complex which also contains biotite- and hornblende-bearing teschenitic basalts and picrites.

In the Glasgow–Paisley area, alkali dolerite sills cut strata ranging from Upper Calciferous Sandstone Measures to Coal Measures. K-Ar mineral dates from four sills are tightly grouped in the range 273 to 270 Ma, suggesting contemporaneity with the Permian sills and lavas of the Ayrshire coalfield. All the major sills are teschenitic, although some contain appreciable nepheline in addition to analcime. A notable example of such a rock is the bekinkinite of Barshaw, near Paisley, a melanocratic theralite with abundant titanaugite and red-brown amphibole (kaersutite). Essexites occur in a small, boss-like intrusion near Lennoxtown.

Four major sill-complexes, some consisting of up to three leaves with a maximum recorded thickness of 36 m, can be traced over wide areas. These occur in the Johnstone–Howwood area; between Paisley and the River Clyde at Scotstoun (the 'Hosie' and 'Hurlet' sills); around Cathcart; and between Necropolis Hill, Glasgow and Easterhouse. Highly altered sills around Milngavie consists of olivine-free dolerite with sporadic, minute patches of quartz. However their mafic minerals (purplish augite, with sporadic red-brown amphibole and biotite) are of the type found in teschenites.

Quartz-dolerite intrusions

Broad, persistent, quartz-dolerite dykes with a general E–W trend cut rocks ranging from Lower Devonian to Middle Coal Measures in the northern part of the Midland Valley (Figure 35). The dykes include fine-grained varieties with a glassy mesostasis which have traditionally been called tholeiites. Thick sills in the Firth of Forth area are closely associated with the dykes in terms of field relationships, petrography and geochemistry. They may be regarded as components of a Midland Valley Sill-Complex (Figure 35), comparable and

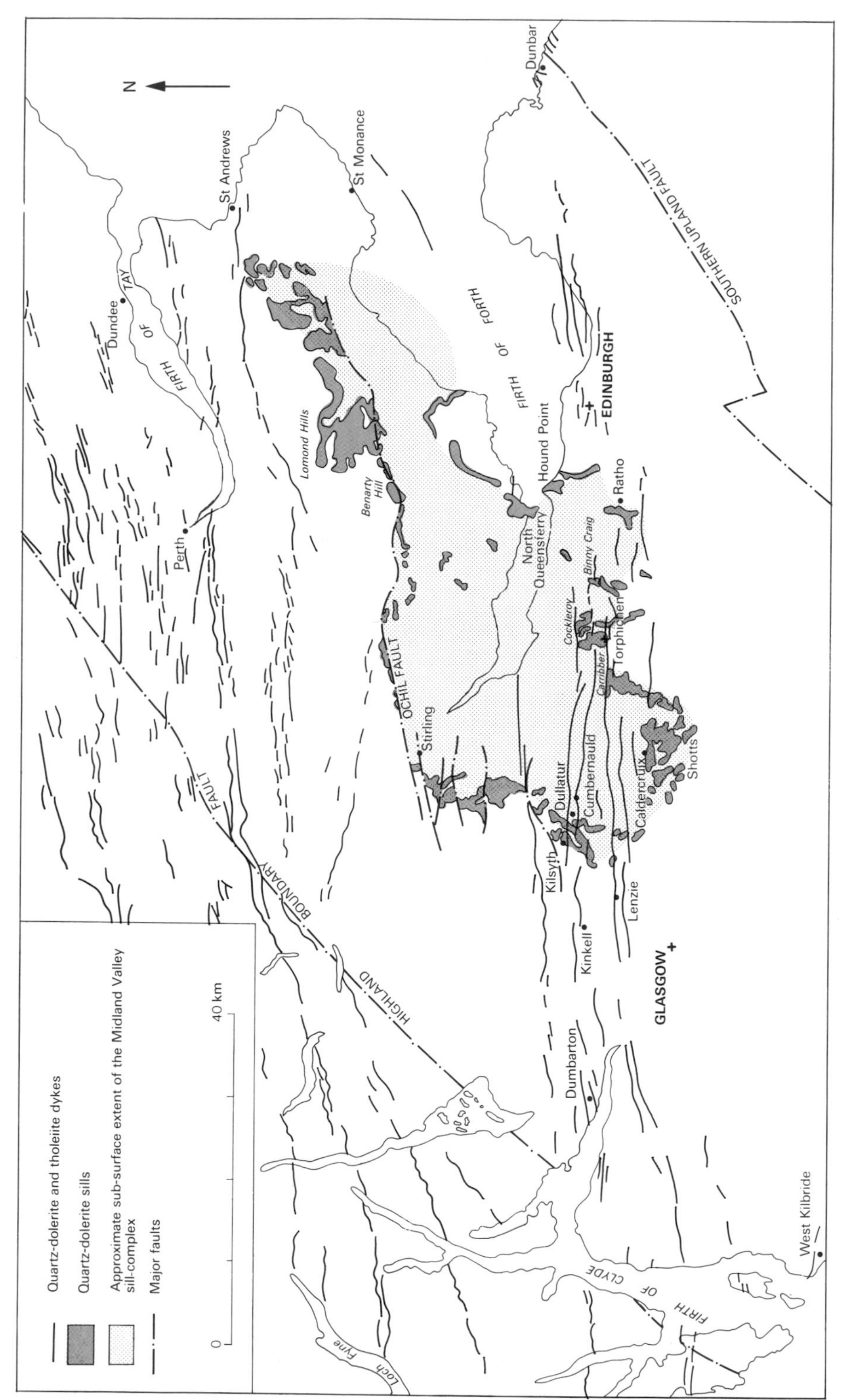

Figure 35 Late-Carboniferous quartz-dolerite and tholeiite intrusions in the Midland Valley

contemporaneous with the Whin Sill of northern England.

Quartz-dolerite intrusions commonly follow E–W fault planes and post-date their main movement. Such faults are known to displace the Coal Measures and in the offshore Fife coalfield quartz-dolerite intrusions cut Westphalian B strata. Field evidence of an upper age limit comes mainly from Fife and is inconclusive: Blocks of quartz-bearing dolerite occur in several vents, most significantly in those of Ardross and St Monance which are considered to be of late-Stephanian age, but also in those of Viewforth and Lundin Links which are of less certain age. Plugs of olivine-basalt and basanite in the Lomond Hills have been considered to cut quartz-dolerite sills but the relationship cannot be proved. Better evidence is found in northern England, where fragments of Whin Sill have been found in a Lower Permian breccia and in the western Highlands, where quartz-dolerite dykes are cut by dykes of the Permian camptonite-monchiquite suite. Whole rock K-Ar dates range from 295 to 290 Ma and a late Westphalian/early Stephanian age for the suite is now generally accepted. The quartz-dolerites therefore represent an interlude of tholeiitic magmatism, which has no known extrusive equivalent, between the major alkaline volcanic episodes of the Dinantian–Namurian and Stephanian–Lower Permian.

Dykes

Many of the dykes cross the Midland Valley in continuous lengths of up to 80 km or may be traced as en-echelon offsets. To the west they cut the Highland Boundary Fault and Dalradian metamorphic rocks and several dykes may be traced almost continuously for 130 km from Loch Fyne to the Tayside area (Figure 35). The more persistent dykes are commonly 20 to 30 m wide with some up to 50 m (Plate 11.1). Geophysical surveys have shown that the swarm continues across the North Sea at least as far as the Central Graben.

In the central Midland Valley persistent quartz-dolerite dykes form a 20 km-wide swarm trending due E–W from Dunbar to Dumbarton. To the south dykes are less numerous and the most southerly major representative occurs at West Kilbride, although many narrower E–W tholeiite dykes in the Ayrshire Coalfield may be of similar age. The extensive dyke swarm north of the Ochil Fault gradually assumes an ENE trend closer to the Highland Boundary Fault. Some dykes are deflected locally into a NE trend in the fault zone.

Sills

Quartz-dolerite sills form prominent scarp features such as the Castle Rock and Abbey Craig, Stirling; the Lomond Hills; and Cockleroy Hill and Carribber Hill in the Bathgate Hills. Other well-known sills crop out at North Queensferry, Hound Point (Plate 10.2), Ratho, the Caldercruix–Shotts area and Kilsyth. Many have been quarried for road metal. Extensive sills are encountered in mines and boreholes in the Central and Fife coalfields where their effect on coal seams is similar to that of the alkali dolerite sills (p.112). Many sills consist of several leaves, 25 to 100 m thick, linked by near-vertical dykes or 'step and stair' transgressions, which are often related to pre-existing fault planes. The whole sill complex is up to 150 m in total thickness and occupies some 1600 km^2.

Francis (1982) has shown that the shape of the sill complex approximates to a series of 'saucers', the lowest and thickest parts of which coincide with the centres of syn-sedimentary Carboniferous basins. In detail, many transgressions can be shown to have occurred in a downward sense, often leaving a thin continuation sill at the higher horizon. It thus seems likely that magma was able to 'flow' down gently-inclined bedding planes into the centres of the basins.

There are no indications of feeder dykes or pipes in the lower, central parts of the complex, but various E–W dykes in the southern part of the regional swarm have long been considered as possible feeders, since some are seen to pass locally into sills (e.g. the Lenzie–Torphichen, Dullatur and Cumbernauld dykes). On the northern margin of the sill complex, the Ochil Fault Intrusion, a series of irregular pods in the plane of the Ochil Fault, is also considered to be connected with the sill-complex and is a possible feeder.

Petrography

A complete spectrum of textures exists between various named types of tholeiite and the quartz-dolerites but mineralogical and geochemical differences are slight. All types occur as broad, persistent dykes, but tholeiites predominate among the thinner dykes and also occur on the margins of thick quartz-dolerite dykes and sills. The textural differences may therefore reflect differing rates of cooling and volatile contents of individual intrusions.

The quartz-dolerites consist essentially of labradorite laths, subophitic augite, pseudomorphs after hypersthene, occasional pigeonite, iron-titanium oxides and an intersertal mesostasis of quartz and alkali-feldspar, usually intergrown as micropegmatite. Amphibole and biotite commonly fringe the augite and oxides. Apatite and pyrite are usual accessories.

Tholeiites are distinguished by the presence of intersertal, pale brown, often devitrified, microlitic glass. Augite is more granular than in the quartz-dolerites, pseudomorphs after olivine occur sporadically, calcium-poor pyroxene is absent and skeletal ilmenite often forms reticulate patterns in the glass. Chlorophaeite (an amorphous mixture of green ferruginous silicates which darkens on exposure to air) is a feature of several tholeiites in which it occurs in intersertal areas or as pseudomorphs after fayalitic olivine.

A fine-grained sill at Binny Craig, West Lothian, consists of a distinctive basalt with small phenocrysts of plagioclase and augite, but is otherwise mineralogically and chemically similar to the quartz-dolerites.

Differentiation of the quartz-dolerites is best observed in the thicker sills and to a lesser extent in wide dykes. Sills are chilled at top and bottom with marginal zones of fine-grained dolerite. The central parts consist of a medium-grained, homogeneous dolerite, in the upper part of which a zone of coarse, irregular, crystallisation commonly occurs. Pink quartzo-feldspathic patches within this zone commonly contain long feathery clusters of augite crystals. Pink aplite veins also occur throughout the sills and quartz-calcite-chlorite veins are abundant in the upper parts. Late stage veins of fine-grained basaltic material, presumably from later pulses of magma, are recorded from both dykes and sills.

Magma genesis and tectonic setting

The magmatism which persisted in the Midland Valley for some 70 Ma during the Carboniferous and early Permian is typical of that which occurs in continental rift environments throughout the world. The alkaline basic rocks range from hypersthene-normative, transitional basalts to more strongly-alkaline, nepheline-normative basanitic and nephelinitic varieties. The rocks of the quartz-dolerite suite have been classed as 'High Fe-Ti tholeiite' type and are intermediate in alkali content between true tholeiitic and alkali basalt magma types.

Recent geochemical studies, mostly by Macdonald and others (1975, 1977, 1980, 1981), suggest that all the magmas have been derived by variable degrees of partial melting of an upper mantle source and that crustal contamination has been negligible. Trace element studies have revealed variations between the basic lavas of different areas and also between individual quartz-dolerite dykes, which have been attributed to long lasting inhomogeneity in the mantle source rocks. Further evidence of the nature of the mantle source has been obtained from mineralogical studies and melting experiments on xenolithic nodules and megacrysts (p.112).

Fractionation of alkali basalt magma at crustal levels to give intermediate and salic lavas appears to have been active only during the Dinantian activity. Almost all of the more fractionated lavas are silica-saturated or oversaturated, although undersaturated, feldspathoidal trachytic rocks occur as intrusions. These differentiates have enabled several magmatic lineages to be recognised from major element chemistry. Field evidence of high-level fractionation processes is seen in several varieties of composite lava flow, which are relatively common in the Dinantian sequences. The tholeiitic magmas probably fractionated at depth from an olivine-tholeiite to quartz-tholeiite and limited high-level fractionation is seen in a few dykes of tholeiitic andesite composition (e.g. the Kinkell dyke, Stirlingshire).

Throughout the period of Carboniferous–Permian magmatism, there is a progressive general tendency for more-alkaline, highly-undersaturated basic rocks to constitute an increasingly higher proportion of the lavas and intrusions. This overall pattern is interrupted by the eruption of transitional or mildly-alkaline basalts, for example in the Passage Group of Ayrshire, and also by the late-Carboniferous tholeiitic intrusions. It is therefore probable that several magmatic cycles occurred throughout the period. Macdonald and others (1977) interpret such cycles as separate thermal events during which large volumes of silica-saturated magmas were generated at high mantle levels, under the high geothermal gradients of the initial stages, followed by smaller volumes of more undersaturated magma from deeper levels as gradients were reduced in the waning stages.

Following the closure of the Iapetus Ocean in Lower Devonian time, the Midland Valley graben began to develop as an intra-continental rift. By Lower Carboniferous time, the area had become part of the southern marginal shelf of the N. America–N. Europe craton and the nearest plate boundaries were well to the south and east of the British Isles. In such a mid-plate environment, tensional stress conditions prevailed and lithospheric stretching led to rifting and increased thermal gradients with consequent mantle melting.

Fundamental crustal fractures with a Caledonian NE–SW trend, which had probably acted as tectono-magmatic controls during the Lower Devonian, continued to influence upper crustal faulting and the siting of volcanic centres during the Dinantian (e.g. the Dumbarton–Fintry line). Such fractures acted as hinge lines, dividing the rift into basins and swells and thereby influencing both volcanicity and sedimentation.

By late-Namurian and Westphalian time, the rift structures had become less active, the structural swells had begun to lose their identity and more widespread sedimentary basins were developed. This coincided with a change towards reduced amounts of generally more-undersaturated and more explosive alkali basaltic volcanicity and the emplacement of alkali dolerite sills within the thickening sedimentary sequences.

During the main phase of the Hercynian Orogeny, plate collision to the south of the British Isles generated dominantly compressive forces in the Midland Valley. Caledonian structures were reactivated and uplift resulted in the absence of Stephanian strata. Subsequent stress release generated new E–W major fractures, extending from Scotland and N England across the North Sea. High geothermal gradients resulted in mantle melting at relatively shallow levels and tholeiitic magmas rose along the E–W fractures.

The igneous activity which followed the tholeiitic episode in the late-Stephanian/early-Permian was entirely of a more-undersaturated nature than any of the preceding episodes and should possibly be considered as a separate tectono-magmatic event. In the eastern Midland Valley volcanicity was still controlled by NE–SW structures (e.g. the Ardross Fault), but in the west, activity seems to have been controlled by new NW- or NNW-trending structures. It has been suggested that these structures are part of a new rift extending from Arran to the Vale of Eden and contemporaneous with similar rifts in the North Sea and Norway. This Permian rift system may signify the initial crustal thinning and break-up of the N America–N Europe craton prior to the formation of a proto-N Atlantic oceanic rift.

13. Tertiary igneous intrusions

Tertiary dolerite dykes, of both tholeiitic and alkali basaltic affinities, with a general NW – SE trend are relatively abundant in the Midland Valley, south-west of a line from Gourock to Douglas (Figure 36). A group of alkali dolerite sills in the Prestwick – Mauchline area are also of Tertiary age.

Dykes

Tertiary dykes of the British Isles occur either in regional linear swarms or in more localised sub-swarms in the vicinity of central intrusive complexes. The majority of the dykes in the Midland Valley are a continuation of a regional swarm, centred upon Mull. On the north-east margin of the swarm a group of wide, persistent dykes traverse the full width of the Midland Valley, extending into the Southern Uplands and some may be traced via intermittent outcrops for over 200 km to the Northumberland coast. To the north-east of this group, Tertiary dykes occur rarely (e.g. east of Glasgow) but to the south-west, many occur sporadically over a width of 30 km. Near the south-west edge of the swarm, the Cumbrae–Stevenston dyke has been traced for almost 300 km to north Yorkshire. It is also one of two dykes which are deflected into an ENE trend for up to 20 km close to the Southern Upland Fault.

The more persistent dykes are commonly 6 to 10 m wide and greater widths of up to 37 m have been recorded from those which extend into northern England. Many impersistent 2 to 3 m-wide dykes, usually of tholeiitic character, are also present. Columnar jointing is typically well developed perpendicular to chilled dyke margins and the rocks are characteristically very fresh and hard (Plate 12.2).

Most of the dykes of the Mull regional swarm are tholeiitic. They are composed of labradorite laths and augite, with varying amounts of glassy mesostasis, often devitrified and darkened by finely disseminated iron-titanium oxides. The larger, more persistent dykes are varieties of tholeiite, quartz-dolerite or tholeiitic andesite which vary considerably in texture along their length. They frequently contain orthopyroxene or pigeonite and generally have a higher silica content than the smaller dykes. The Cumbrae–Stevenston dyke is the most siliceous of this group and is characterised by numerous large phenocrysts of anorthite-bytownite set in an acidic, glassy groundmass constituting the rock type cumbraeite.

Most of the NW–SE and NNW–SSE-trending dykes exposed along the Clyde coast from Troon to beyond Girvan probably belong to a separate Arran sub-swarm or to continuations of regional swarms passing through Jura and Islay. The majority of these dykes are alkali olivine-dolerites, but tholeiitic dykes are also present. Most of the olivine-dolerites contain small amounts of analcime or zeolites and consist essentially of ophitic purplish titanaugite enclosing labradorite laths, olivine and iron-titanium oxides. Such rocks are similar to finer-grained dykes and sills of Kintyre and Arran which have been termed crinanites.

1 Dune bedding in the Permian Mauchline Sandstone, Mauchline. Quarry is now filled in. (C 2913)

2 Lion Rock, Great Cumbrae. Tertiary age tholeiitic andesite dyke of cumbraeite type intruded into Upper Devonian sandstones. Former sea stack now preserved on post-glacial raised beach. (D 3530)

Plate 12

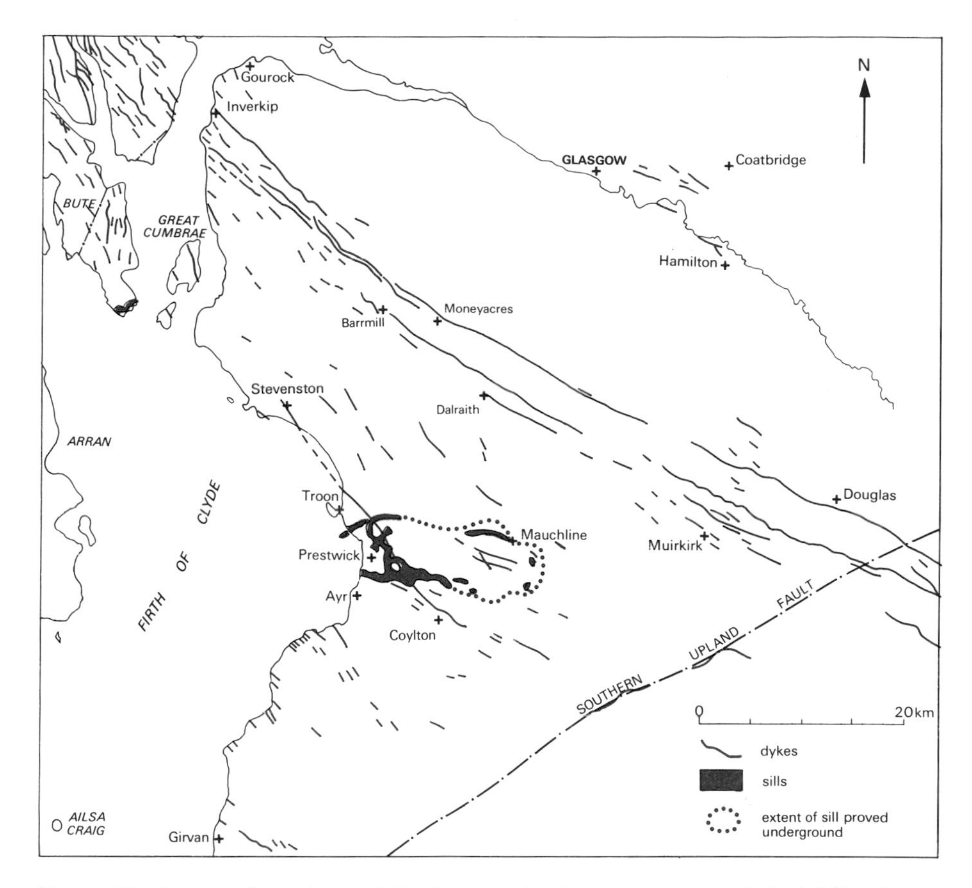

Figure 36 Igneous intrusions of Tertiary age in the western part of the Midland Valley

Members of the dyke swarms cut post-Coal Measures alkali dolerite sills, E–W quartz-dolerite dykes, Permian volcanic necks, the Permian Mauchline Sandstone and the Triassic sandstones of Arran. The contemporaneous Cleveland Dyke of north Yorkshire cuts Jurassic sediments and the obvious association of the swarms with the Mull and Arran central complexes leaves the Tertiary age in no doubt. Evidence from Arran indicates that the alkali dolerite (crinanitic) dykes and sills are earlier than the Northern Granite and Central Ring Complex, all of which are cut by tholeiitic dykes.

Sills

Several extensive outcrops in the Mauchline basin of ophitic, analcime-bearing olivine-dolerite (crinanite), cut by analcime-syenite veins, are continuous at depth and form the Prestwick–Mauchline Sill-Complex. The complex has a maximum thickness of 60 m and is believed to consist of between one and three leaves, up to 500 m apart, connected by dykes or steeply inclined sheets. Several transgressions elevate the sill from the top of the Middle Coal Measures in the west, through the Upper Coal Measures, to the Mauchline Volcanic Group in the east and the Mauchline Sandstone in the north-east.

The petrographic similarity to the Tertiary crinanite sills of Arran and to some of the thicker Tertiary dykes of Ayrshire suggests a Tertiary age and this has been supported by palaeomagnetic measurements and a K-Ar mineral date of 57 ±1.4 Ma.

Magma genesis and tectonic setting

Recent unpublished investigations reported by Thompson (1982a, p. 470) reveal that most of the dykes in the Midland Valley of the Mull regional swarm are magnesium-poor basalts and tholeiitic andesites, corresponding to varieties of the 'Non-Porphyritic Central' magma type first defined from Mull and subsequently recognised in the NE England dykes. Significant numbers of the tholeiitic dykes have the diagnostic chemistry of the magnesium-rich, low-alkali 'Preshal Mhor' magma type of Skye, which is close in composition to 'Mid Ocean Ridge Basalt' and is now widely recognised in the British Tertiary province. Alkali olivine-basalts of the Mull 'Plateau' magma type are represented by the crinanites and allied rocks which are particularly abundant in the Arran sub-swarm.

Current theories summarised by Thompson (1982a,b) suggest that the alkali 'Plateau' magmas were generated by partial melting of mantle material at the base of the lithosphere. The tholeiitic 'Preshal Mhor' magmas were probably generated by further melting of upper mantle already depleted by earlier melting episodes and may have evolved the 'Non-Porphyritic Central' magmas, which occupy so many of the regional dykes, by fractionation in the upper crust.

The persistent NW–SE trend of the regional dyke swarm indicates that it is controlled by fundamental deep crustal fractures. It has been suggested that such fractures may have been in existence throughout the late-Palaeozoic and Mesozoic, when they exerted control on sedimentary basins and igneous activity and possibly acted as channels for mineralising solutions (p.152). The regional dykes most likely originated as vertical intrusions from ridge-like magma chambers aligned along the fractures.

Palaeocene igneous activity in the British Isles was concentrated over the period between 60 and 52 Ma. The regional dyke swarms were probably intruded during the later part of this activity which is thought to be coincident with the initiation of the final stage of opening of the NE Atlantic between Greenland and NW Europe. It may be that the dyke swarms follow rift-normal tensional fractures associated with the spreading, which extend well into the continental margin.

14. Structure

The Midland Valley is in structural terms a graben, and is approximately 80 km wide, bounded in the north by the Highland Boundary Fault and in the south by the Southern Upland Fault. The graben structure developed as such in the Devonian, but its location and dimensions were determined by crustal events which happened in the Lower Palaeozoic.

The Grampian Highlands are the result of intense deformation and metamorphism in early Ordovician times of a thick pile of Dalradian sediments. The Dalradian sediments accumulated on the southern part of a northern continent which included North America, Greenland and Scandinavia. The Iapetus or Proto-Atlantic ocean lay to the south of this northern continent.

The Southern Uplands rocks are thought to have accumulated as a result of subduction of oceanic crust beneath the North American–North European continent. Lower Palaeozoic sequences of basalt, chert and thick greywackes accumulated on the floor of the Iapetus Ocean and were progressively scraped off as the oceanic plate moved north-westwards into the subduction zone. The sediments accumulated and were deformed as successively younger slices were thrust under each other to form an imbricate accretionary prism. The accretionary prism became emergent in the northern part of the Southern Uplands and shed detritus into the southern part of the Midland Valley in Silurian times.

The final closure of the Iapetus Ocean and the cessation of subduction is assumed to have occurred in the late Silurian or Devonian. Several plate-tectonic models for the Caledonian orogeny have been proposed.

The deep structure of the Midland Valley is poorly known and the evidence is fragmentary, but it appears to be fundamentally different from the Grampians to the north and the Southern Uplands to the south. It is separated from these two areas by major crustal lineaments which date from the Lower Palaeozoic and which more or less pre-determined the location of the lines of weakness which define the graben.

The positions of the major faults and folds in the region are shown on Figure 37.

Faulting

Highland Boundary Fault

The Highland Boundary Fault extends from Stonehaven in the north-east to near Helensburgh on the Firth of Clyde. It separates the little-deformed rocks of the Midland Valley from the intensely deformed rocks of the Highlands.

The fault is only one of a number of parallel fractures which form a zone usually consisting of two or three major and several minor fractures, with the Highland Boundary Fault itself as the most south-easterly of the major faults.

The inclination of the major faults, as seen in exposure, is steep towards the

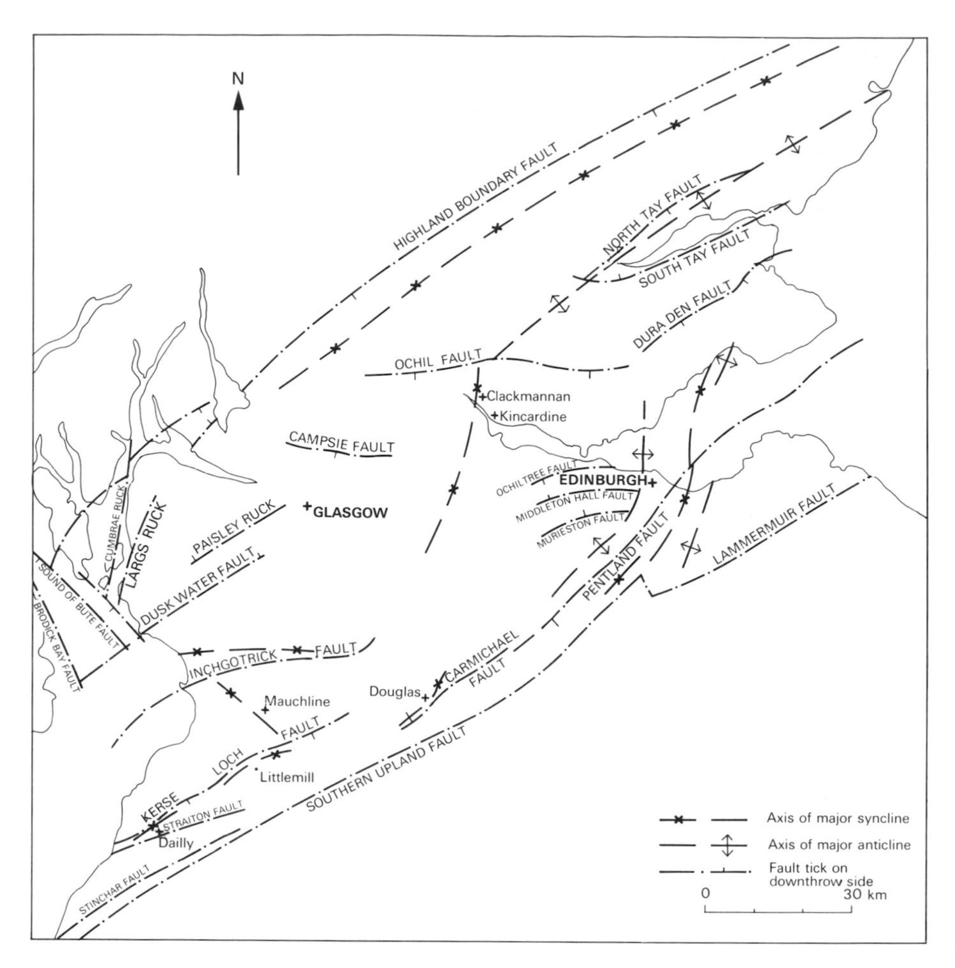

Figure 37 Principal faults and folds in the Midland Valley

north-west between Stonehaven and Aberfoyle, but from Loch Lomond to the Clyde coast its hade is steep towards the south-east. The apparently straightforward south-easterly downthrow between Stonehaven and Aberfoyle is complicated by later movements affecting younger sediments in the Loch Lomond to Firth of Clyde area. The fault bifurcates south-west of Loch Lomond into a northern branch which passes north-west of Helensburgh and crosses the Rosneath Peninsula, Toward Point and the Island of Bute. The southern branch passes south-east of Helensburgh and its course is obscured south of the Clyde by the cover of Carboniferous rocks.

The line of the fault is marked in places by fault slices of a variety of rocks of the Highland Border Complex, which is thought to be of Lower Ordovician age. The rocks include serpentinite, cherts, spilites and limestone.

The main displacement on the fault occurred prior to deposition of the Upper Devonian and probably occurred both during and after deposition of Lower Devonian. The amount of downthrow to the south-east has been estimated to be of the order of 2500 to 3000 m in the Crieff and Tayside area. Evidence of downthrow to the north-west, after deposition of the Upper

Devonian and basal Carboniferous sediments, is preserved in the Loch Lomond to Firth of Clyde area. Gravity measurements suggest that the Upper Devonian north-west of the fault is underlain by several hundred metres of Lower Devonian sediments.

The possibility that transcurrent movement has taken place on the fault has been discussed by Anderson (1947) and it is inferred to have occurred in the Ordovician according to the plate-tectonic model of Lambert and McKerrow (1976). Bluck (1980) also suggested lateral movement on the fault during Devonian times, but in a sinistral sense, opposite to that of Lambert and McKerrow. No such movement has yet been proved, although sinistral displacement is known to have occurred on faults with NNE trends in the Highlands and these faults trail into the Highland Boundary Fault.

The north-westerly hade of the fault, in exposure, is supported by interpretation of the gravity measurements which suggest that the hade is between 10° and 20° to the north-west, at least as far as the base of the Lower Devonian. This evidence suggests that the fault is a high angle reverse fault. This is likely to be too simple a description, but stratigraphic evidence enabling a more detailed analysis is lacking.

Southern Upland Fault

The Southern Upland Fault runs from Glen App in the south-west to Leadburn in Midlothian where the fault apparently is buried by the cover of Carboniferous sediments. It is perhaps continued *en echelon* by either the Lammermuir Fault which runs from south of Leadburn to the coast near Dunbar or the Pentland Fault which separates the Pentland Hills from the Midlothian basin.

The Southern Upland Fault and the Lammermuir Fault separate the steeply dipping, folded and faulted rocks of the Southern Uplands from the more gently deformed strata in the Midland Valley. In the south-west, in the Glen App area, the fault lies within the outcrop of the Lower Palaeozoic rocks and the line of separation between the Southern Uplands and the Midland Valley, in terms of the sediments and their deformation, is that section of the Straiton Fault which lies between Girvan and Dailly.

The Southern Upland Fault is accompanied by a group of associated faults subparallel to the main fault, including the Straiton Fault, the Kerse Loch Fault, the Carmichael Fault and the Pentland Fault among others.

Displacement began at least as early as Mid-Devonian and was renewed, in some instances with the throw reversed, during the Carboniferous and later. The main displacement was probably in the Lower and Middle Devonian with a net downthrow to the north-west across the zone, and it occurred along the whole length of the fault. Later movements, during the Carboniferous, were more local in their effect. Differential subsidence across the lines of faulting resulted in abrupt changes of thickness in the Carboniferous of Ayrshire and south Lanarkshire, but similar variations in thickness in relation to faulting are absent in the Lothians.

In several instances later movements on a fault in the zone are in an opposite sense to the earlier displacement. The relationships of the rocks on either side of the Straiton Fault indicate that downthrow to the south-east occurred prior to deposition of the Upper Devonian, but the direction of displacement was reversed later causing Carboniferous and Upper Devonian rocks, resting unconformably on Silurian rocks, to be downfaulted against Lower Devonian rocks to the south-east. The Southern Upland Fault between New Cumnock and Sanquhar is a double fault with a narrow strip of Ordovician rocks in faulted relationship to Lower Devonian rocks in the north-west and Coal Measures sediments to the south-east. The displacement on the northern fault occurred before Carboniferous deposition and that on the southern fault is post-Carboniferous. The pre-Carboniferous throw has been estimated to be about 900 m. A similar reversal of throw occurs on the Pentland Fault.

The net displacement on the fault zone is unknown. The several components of the displacement on each fault vary in amount and in direction from place to place. Only locally can a component of the throw be measured where it affects Carboniferous rocks.

The Southern Upland Fault is assumed to be primarily a normal fault, but transcurrent movement has been suggested to explain the variability of the downthrow.

The Pentland Fault, in its post-Carboniferous displacement, is a reverse fault. The hade has been measured at 22° to the north-west by drilling and the north-west limb of the Midlothian syncline is locally overturned. The fault can be traced offshore in the Firth of Forth and it passes between the east Fife coast and the Isle of May.

Other faults

Other faults with a trend more or less parallel to the marginal faults include the Dusk Water Fault and the Paisley Ruck. Both faults were active during Carboniferous deposition and the latter has a downthrow to the north-west of up to 500 m.

In the north-east of the region the North Tay, South Tay and Dura Den faults all trend approximately north-east. The North Tay and South Tay faults form a graben structure on the crest of the Sidlaw Anticline and have throws of more than 500 m and about 1 km respectively. Upper Devonian and Carboniferous sediments are preserved in the graben. The Dura Den Fault has a downthrow to the south of not less than 300 m.

The Ochil, Campsie and the Inchgotrick faults are major fractures, oblique to the marginal faults. The Ochil Fault, like the others, is a normal fault and has a maximum downthrow of about 3000 m near Alva. Most of the throw is post-Carboniferous, but there may have been movement during Carboniferous sedimentation.

Two important NNE-trending fold-fault structures occur in the Firth of Clyde area. The Cumbrae Ruck traverses the Great Cumbrae and geophysical evidence suggests that it causes a sinistral displacement of the Highland Boundary Fault off Cloch Point, near Gourock. A parallel structure crosses the Hunterston peninsula and runs north-north-east into the Renfrewshire Hills behind Largs. Downthrow is on the eastern side of the faults except in Cumbrae where Lower Carboniferous strata are downthrown on the west side.

The pattern of faulting revealed by mining information in the coalfields shows two predominant trends of normal faulting. A W–E trend appears to be earlier since the other, NW–SE trend, terminates against or trails into the W–E trend (Anderson, 1951). Faults with a W–E trend are particularly numerous and include several with a considerable downthrow. In West Lothian there are three important W–E faults. The Ochiltree Fault has a downthrow of about 360 m on the south side. The Middleton Hall Fault and the Murieston Fault both throw down on the north side, with a displacement on the former of about 470 m and on the latter of about 550 m. Faults trending W–E are particularly numerous in a zone between the Clyde and the Forth.

Three sub-parallel NW-trending faults are known from geophysical evidence in the Firth of Clyde. The Ardrossan Fault and the Sound of Bute Fault both throw down to the south-west and limit the North-East Arran Trough on its north-east side. The south-west side of the trough is bounded by the Brodick Bay Fault.

Folding

The Lower Palaeozoic rocks of the Straiton area and the Pentland Hills show evidence of folding prior to deposition of the Lower Devonian. The rocks are steeply inclined or vertical with a NE strike and are unconformably overlain by Lower Devonian strata. However, in the Lesmahagow and Hagshaw Hills inliers, and at Stonehaven, Lower Devonian rocks follow Silurian rocks with apparent conformity.

The Lower Devonian rocks are folded about NE-trending axes and the folds are tightest, with locally overturned limbs, in zones adjacent to the marginal faults. Away from the north and south margins the folds are more open, but of considerable amplitude. In the north-east of the area the NW limb of the Strathmore syncline is locally overturned in the vicinity of the Highland Boundary Fault, but the Sidlaw anticline is a broad open structure.

The main structures in the Carboniferous in the western part of the Midland Valley differ in style and axial trend from those in the eastern part. In the west there are broad regional warps with a north-westerly or westerly trend and locally north-easterly trending folds associated with faults, but in the east the main axes run a few degrees east of north.

In the west the Mauchline basin has a broad synclinal structure with a north-westerly orientation. The SW limb of the syncline appears to culminate in the Kirkoswald area. To the north of the Mauchline basin the axis of maximum thickness of the Clyde Plateau lavas, which runs from Greenock to Strathaven, is the primary control on the outcrop and the broad W–E syncline in the Kilmarnock and Irvine area is truncated on the south side by the Inchgotrick Fault.

Folds with a north-easterly trend are associated with faults of the same orientation and occur on the downthrown side. These folds are in part contemporaneous with Carboniferous sedimentation and are exemplified by the Dailly, Littlemill and Douglas synclines.

In the east the main structures are the synclines which contain the Central and Stirling and Clackmannan coalfields and the Midlothian Coalfield. These structures trend a few degrees east of north. The axis of the Clackmannan Syncline, in the Stirling and Clackmannan Coalfield is more or less coincident with the Kincardine basin. The syncline was forming during deposition of the Upper Carboniferous and possibly also in Lower Carboniferous times. The formation of the syncline is thought to have been complete by the end of the Carboniferous.

The main structure affecting the Carboniferous rocks in Midlothian and Fife is comparable to the Clackmannan syncline, but the relationships are obscured by later displacement on the Pentland Fault. The basin is parallel to the Clackmannan syncline and isopachytes in the Carboniferous of Midlothian indicate thickening of Dinantian and Namurian sediments towards the axis of the structure from the eastern side. The western half of the depositional basin in Midlothian is cut out by post-Carboniferous displacement on the Pentland Fault.

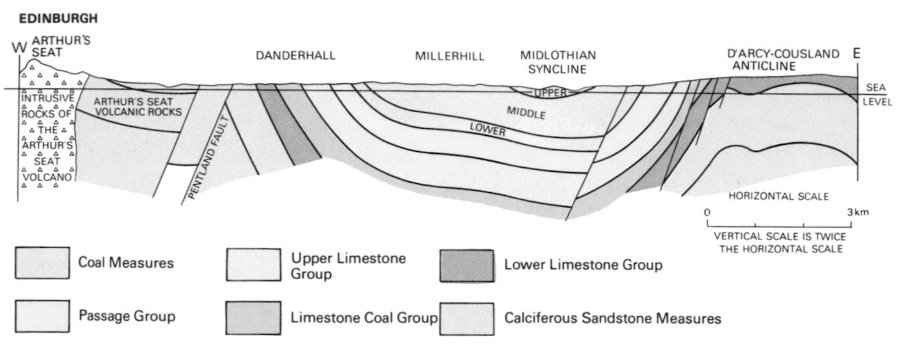

Figure 38 Cross-section of the Midlothian Syncline

The syncline in Midlothian is markedly asymmetrical (Figure 38). The NW limb is highly inclined and has in places subsidiary anticlines. The strata have dips in the range 60° to 80° and locally they are overturned. The syncline is flanked on the east side by the D'Arcy–Cousland anticline which brings to outcrop a core of Dinantian rocks flanked by Namurian. Both syncline and anticline close to the south-south-west.

15. Quaternary

The Quaternary period extends from about 2 Ma to the present day and it is characterised by considerable and repeated climatic fluctuations. Temperate episodes during which conditions were similar to, or warmer than, the present day alternated with episodes of arctic conditions. During the colder glacial periods, glaciers and ice-sheets formed at relatively low latitudes in areas of sufficient precipitation. At its maximum extent the ice-sheet which covered much of northern Europe and the British Isles extended as far south as the Thames. Stratigraphical and biological evidence for several episodes of glaciation has been found at several sites in East Anglia and the Midlands of England. The stratigraphical subdivison of the Quaternary reflects the climatic fluctuations and the stages of the Quaternary represent alternating cold and temperate episodes. Indications of climatic conditions are given by pollen analyses, aquatic faunas and a study of insect assemblages, particularly Coleoptera (beetles). The latter have a much more rapid response to climatic change than does the tree cover.

The evidence of a sequence of glacial and interglacial episodes in England implies that Scotland also underwent several glaciations during the Quaternary. Interglacial sites in Scotland are few and in the Midland Valley there is little evidence of events prior to the last glaciation. With the exception of one or two instances, older Quaternary deposits have either not been recognised or have been obliterated by the effects of the last glaciation. Only the two youngest stages of the Quaternary are known. The Devensian Stage includes the glacial and fluvioglacial deposits, and the Flandrian Stage is represented by the post-glacial deposits. The boundary between the Devensian and Flandrian Stages is taken at 10 000 years B.P. (radiocarbon years before present).

Devensian

Very little evidence exists of Devensian events prior to the retreat of the last glaciation, but biological remains at three sites give an inkling of conditions earlier in the Devensian. Bones and a tooth of the woolly rhinoceros, *Coelodonta antiquitatis* have been found in sands and gravels beneath till in the Kelvin Valley near Bishopbriggs. One of the bones gave a radiocarbon date of 27 500 (+ 1370, − 1680) years B.P.

In the early part of the 19th century animal remains including mammoth tusks, reindeer antlers and arctic shells were found in a pit near Kilmaurs. The fossils were obtained from beds of sand and clay beneath till. A mammoth tusk gave a radiocarbon date of 13 700 (+ 1300, − 1700) years B.P., but a reindeer antler believed to come from the same deposits was found to have a radiocarbon date of >40 000 years B.P.

Since it is possible that one or more of the dated fossils from Kilmaurs and Bishopbriggs have been derived from pre-existing sediments the stratigraphic significance of the material is limited. In addition doubt has been cast on the validity of the radiocarbon dates (Sissons, 1981) and until either of the finite dates can be confirmed conclusions drawn about the limits of the ice-sheet at those dates must be tentative.

At Burn of Benholm, near Johnshaven in Kincardineshire lenses of peat intercalated in the basal part of the till have been dated by radiocarbon measurements at >42 000 years B.P. The pollen analyses of the samples indicate a tundra environment and the peat is interpreted as being of Early or Middle Devensian age incorporated in a Late Devensian till. At some time during the Devensian glaciation Scandinavian erratics were transported across the North Sea, perhaps indirectly, and were deposited in NE Scotland.

Glaciation

The last glaciation to affect the Midland Valley occurred in the latter part of the Devensian Stage. The ice-sheet spread into the Midland Valley from centres in the western part of the Grampian Highlands and to a lesser extent from the Southern Uplands. The direction of ice movement is shown in Figure 39. The ice-sheet is thought to have reached its maximum development about 18 000 years B.P. when the entire region was buried beneath a thick cover of

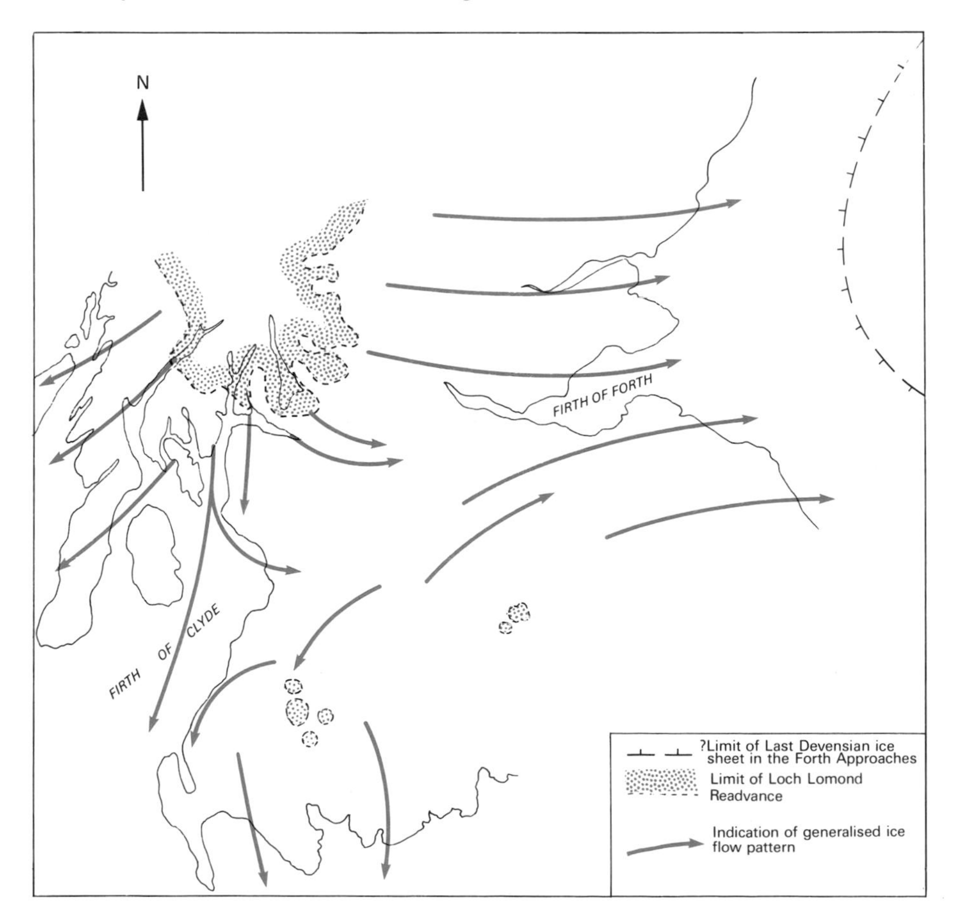

Figure 39 Generalised pattern of ice-flow and ice limits

ice. Estimates of the maximum height of the ice-sheet surface are of the order of 1500 to 1800 m.

Towards the end of the Devensian a period of climatic amelioration caused the ice-sheet to retreat and vegetation colonised the exposed surface. Radiocarbon dates indicate that the Midland Valley was free of ice shortly after 13 000 years B.P., but a deterioration in the climate allowed glaciers to form again in the Grampians and locally in the western part of the Southern Uplands. Glacial encroachment into the Midland Valley occurred in the upper parts of the Teith and Forth valleys and south-east of Loch Lomond (Figure 39). This episode is known as the Loch Lomond Readvance and it occurred between about 11 000 and 10 300 years B.P.

The climatic conditions between the glacial retreat and the Loch Lomond Readvance, according to the evidence of the Coleoptera, are thought to have been comparable to the present day between 13 000 and 12 000 years B.P., but deteriorated thereafter. The interval is known as the Windermere or Late-Glacial Interstadial and the episode which includes the Loch Lomond Readvance is referred to as the Loch Lomond Stadial.

Other readvances of the ice have been postulated by various authors but the Loch Lomond Readvance is at present the only generally accepted major glacial readvance known to have affected the Midland Valley.

A rapid improvement in the climate at the end of the Devensian brought glaciation to an end.

Glacial erosion

The movement of ice across the landscape caused considerable modification to the topography and the most intense erosion occurred in the valleys in the vicinity of the ice dispersion centres in the Highlands. Glacial erosion features in the Midland Valley are numerous but usually less dramatic. However, where there were constrictions or obstructions to ice movement the amount of erosion was considerable. The glaciated landforms may relate to more than one glaciation.

Ice moved over the entire area causing a greater or lesser amount of steamlining or gouging according to the resistance of the underlying rock and the power of the ice at any particular point. Ice flowed into the valleys in the lowlands, modifying their outline, overdeepening them and spilling through breaches in the watershed into adjacent valleys. Glacial breaches occur in the Ochil Hills and in other areas of high ground. Crag-and-tail features are numerous and glacially eroded depressions around the upstream end of these features indicate the scale of erosion. It is estimated that at least 105 and 150 m of rock have been removed from around Edinburgh's Castle Rock and Salisbury Craigs respectively. Crag-and-tail features and other elongate ice-moulded landforms help to indicate directions of ice movement. Less obvious but very considerable erosional effects are revealed by boreholes which indicate the presence of rock basins and glacially overdeepened valleys concealed by superficial deposits. Closed channels in the rock surface beneath the Devon valley are more than 100 m below OD. Similarly a buried trench in solid rock in the valley of the Forth west of Queensferry is known to be up to 200 m deep. In the area west of Stirling it has been estimated that over 100 m of Devonian sediments have been removed over a large area by glacial erosion.

During deglaciation meltwater rivers within, below or at the margins of the ice, eroded channels in the ice and in the underlying till or rock. The channels commonly run obliquely downhill and are often discontinuous. They also occur cutting valley spurs. The channels range in size from 1 to 2 m to over 20 m deep and may be occupied by misfit streams whereas others are dry. Such channels are common in many parts of the region and have been described from several areas.

Glacial deposits

Till

The detritus of glacial erosion was deposited as till either below the ice-sheet as lodgement till or laid down from melting ice as ablation till. The former is of wide distribution and the latter is generally thin. The composition and colour of the till depends to a large extent on the lithology of the underlying bedrock but it also contains rock fragments carried from a distance. Till derived from Devonian sediments has a reddish brown colour and contains a higher proportion of sand than till which consists predominantly of the detritus from Carboniferous sediments. The latter is usually dark brownish grey and contains a higher proportion of clay and silt. Undisturbed lodgement till is commonly a very firm, tough deposit.

The till forms a mantle of irregular thickness being generally thickest in low ground and patchy or absent on the hill tops. The more resistant rock types which have suffered less erosion tend to project through the mantle of till to form crags or rock knobs. Drumlins, elongate hills formed of till, some with a rock core, are common in the low ground in the Clyde and Forth valleys and in Ayrshire.

In the North Sea deposits of till and marine sediments, known as the Wee Bankie Beds, are believed to be the terminal moraine of the last ice-sheet and the eastern limit of these beds indicates the approximate maximum extent of the late Devensian glaciation.

In several areas there is more than one till. Three can be distinguished in Midlothian with fluvioglacial sand and gravel deposits between them. The content of the tills indicates that the basal one was deposited by ice moving eastwards. The overlying red-brown till was deposited by ice from the Southern Uplands and the top one, the Roslin Till, which has a limited distribution, is a very local deposit. There are no organic remains in the beds between the tills which could be used for dating.

In the Glasgow area a red till overlies a grey till, but this is thought to be related to different bedrock sources and that the two tills were deposited virtually contemporaneously.

Trails of erratic blocks carried by the ice downstream from readily identifiable sources are well-known in the Midland Valley. The most familiar is the trail of boulders of essexite extending eastwards from the small essexite plug near Lennoxtown. The trail is at least 64 km long. Other examples include the transport of granite fragments into the Midland Valley from the western part of the Grampians and from the granites of the western Southern Uplands.

There are several very large masses of rock which are thought to have been transported by ice. The largest is a mass of limestone near Kidlaw in East Lothian which is about 530 m long by 400 m wide and was formerly quarried. The mass is thought to have been moved a few kilometres to the east.

1 Exhumed, ice-scoured boulder of Devonian conglomerate in outwash gravels of Loch Lomond Readvance, near Callander. (D 3326)

Plate 13

2 Frost wedge in cross-bedded glacial sand, Drumclog, Lanarkshire. (D 2305)

Plate 14 Quaternary fossils

1 *Chlamys islandica*, ×0.5. **2** *Balanus porcatus*, ×0.7. **3** *Laevicardium crassum*, ×0.7. **4** *Nuculana pernula*, ×2. **5** *Buccinum undatum*, ×0.7. **6** *Macoma calcarea*. **7** *Lucinoma borealis*. **8** *Ophiolepis gracilis*. **9** *Corbula gibba*, ×2. **10** *Trophonopsis clathratus*. **11** *Puncturella noachina*, ×2. **12** *Lymnaea peregra*, ×2. **13** *Mya truncata*. **14** *Littorina obtusata*, ×2.

In several areas the till contains marine fossils, including bivalves and foraminifera. Shelly till covers much of the low ground in Ayrshire extending up to a height of at least 300 m OD in eastern Ayrshire, and till containing foraminifera occurs in the Glasgow area. Ice moving from the Highlands picked up marine organisms, either from the sea bed or from an area formerly covered by the sea, and spread into the Glasgow area and central Ayrshire. Shells dating from the Late-Glacial Interstadial, transported by ice during the Loch Lomond Readvance have been found in moraine and fluvioglacial deposits near the Lake of Menteith and near Drymen.

Fluvioglacial deposits

When the climate improved, towards the end of the Devensian, the ice melted and water flowed on, through and under the ice carrying glacially eroded debris. The sediment was deposited either within, or in contact with, the ice as kames and eskers or was carried beyond the ice-front and laid down in broad outwash fans. The finer fraction consisting of clay and silt was carried farther and was deposited in lakes or in the sea. In the last stages of melting, areas of inactive ice, surrounded and partially covered by sand and gravel, tended to block the drainage. Temporary ponding of water allowed the deposition of silt and clay.

There are extensive fluvioglacial deposits in Strathmore, in parts of Fife and Kinross and also in a belt along the south side of the Midland Valley between Dunbar and Lanark. Smaller areas of sand and gravel also occur around Darvel, Drymen and in the valley of the River Kelvin.

Large areas of glacial outwash sand and gravel occur in Strathmore where meltwater issued from the Highland valleys. Examples of these are found near Blairgowrie and Edzell. The deposits are coarser near the Highland border, where they consist of coarse, cobble gravel, and they pass into sands and finer gravels farther away.

The area of mounds and ridges of sand and gravel around Carstairs is an example of a system of eskers. They were formerly believed to be moraine marking the limit of the Highland ice or the Southern Uplands ice. The landforms consist of a series of branching ridges up to 30 m high enclosing numerous kettleholes.

Marine sediments

Marine sediments of late Devensian age occur commonly in the Tay and Forth estuaries and at various sites around the Firth of Clyde. They are late-Glacial in age and have given radiocarbon dates up to about 13 780 ± 120 years B.P. The sediments consist of laminated silt and clay, shelly clays, silts and silty sands, overlain by sands and gravels. They contain a fauna of molluscs, foraminifera and ostracods. The coarser sediments are, in part at least, reworked fluvioglacial deposits. Their occurrence on land is due to a subsequent relative drop in sea level.

In the Firth of Clyde area the 'Clyde Beds' have radiocarbon dates between 13 780 ± 120 years B.P. and about 10 200 years B.P. In the Tay–Earn area, the late-Glacial marine clays are subdivided into the Errol Beds which were deposited prior to 13 500 years B.P. and the younger Powgavie Clay which contains a fauna similar to that of the Clyde Beds in the west. The Errol Beds were deposited during deglaciation and the Powgavie Clay after deglaciation.

The upper part of the clay sequence is replaced west of Errol by the marine sands of the Culfargie Beds.

In the North Sea the Marr Bank Beds are shallow water sands with a cold water fauna. They are thought to be contemporaneous with the Wee Bankie Beds and were deposited in the sea to the east of the ice limit (Figure 39).

Late Devensian and Flandrian fossil shorelines

Raised beaches are a prominent feature of the coastal landscape in Scotland and are a consequence of changes in the relative position of sea level during the Quaternary. The position of sea level varied in relation to the interaction of two main causes. Eustatic lowering or raising of sea level occurred in response to the formation or wasting of vast ice-sheets, and isostatic uplift occurred when the glaciated area was relieved by melting of the mass of ice. The isostatic uplift was centred in the western Grampian Highlands and the amount of uplift diminished radially from the centre. The differential isostatic recovery after glaciation had the effect that the oldest beach now has the greatest gradient outwards from the centre of isostatic uplift and the gradient diminishes in successively younger shoreline features.

During deglaciation the sea level fell from a level relatively higher than at the present time and a number of shoreline features in east Fife, and in Angus and Kincardineshire have been interpreted as late Devensian shorelines formed in association with a westerly retreating ice margin (Cullingford and Smith, 1966, 1980). The younger, lower, shoreline features have lower gradients than their predecessors and extend farther to the west (Figure 40). However the correlation of individual shoreline fragments in east Fife and their relationship to actual shorelines of the period has been questioned (Forsyth and Chisholm,

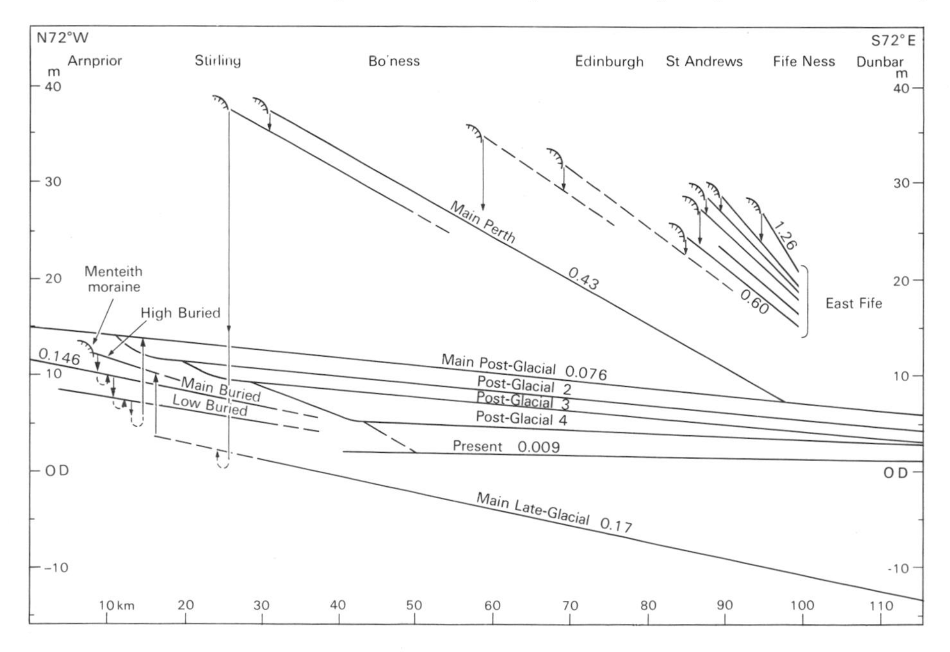

Figure 40 Sequence of shorelines in SE Scotland with gradients in m/km (from Sissons, 1974)

1 Post-glacial raised beach deposits on rock platform of Upper Devonian rocks. Prominent cliff marks line of former coastline, south of West Kilbride, Firth of Clyde coast. (MNS 1892)

Plate 15

2 Hyndford sand and gravel quarry, near Lanark. Cobble and boulder gravel overlying sand and gravel. (D 2610)

1977). The most obvious late-Glacial shoreline is the Main Perth Shoreline which has its western limits in outwash deposits between Falkirk and Stirling.

Continued emergence led to a period of relatively low sea level during which the Main Late-Glacial Shoreline was formed. In the east around Grangemouth the shoreline is represented by a layer of gravel buried by later deposits resting on a bevelled surface of rock, till and late-Glacial marine deposits. This shoreline slopes gently eastwards and passes below Ordnance Datum near Grangemouth (Figure 41). In the Firth of Clyde a prominent rock platform and cliff is correlated with the Main Late-Glacial Shoreline. The feature has a gradient to the south and south-west and it passes below Ordnance Datum in south Kintyre. The age of the raised beach in the west of Scotland was orginally believed to be Post-Glacial but subsequently an interglacial age became accepted. More recently it has been suggested that it was formed more or less contemporaneously with the Loch Lomond Readvance. The relatively short period available for the formation of a very prominent rock platform and cliff is explained by assuming particularly intense frost shattering in the intertidal and splash zones.

A relative rise in sea level following the formation of the Main Late-Glacial Shoreline and a subsequent intermittent fall caused the formation of the High, Main and Low Buried Beaches. These features are known from borehole information in the Carse of Stirling. Stepped features overlain by thin peat layers are interpreted as beaches that have been buried under the Carse Clay during the Post-Glacial transgression. The age of the High Buried Beach is taken as 10 300 to 10 100 years B.P. and is believed to be contemporaneous

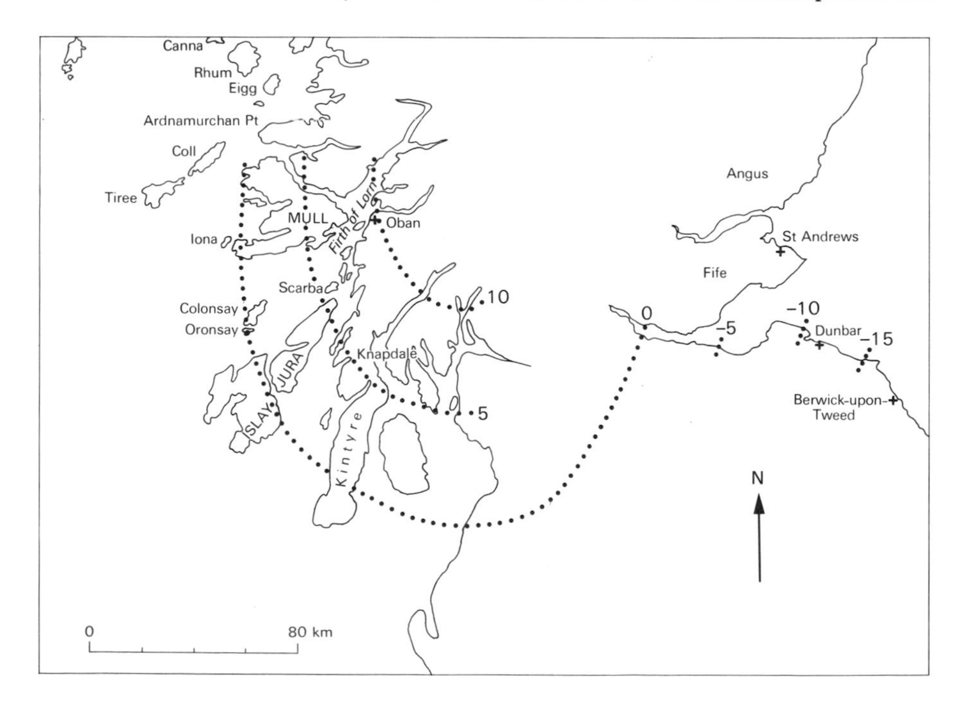

Figure 41 Isobases (in metres) of the Main Late-Glacial Shoreline (after Dawson, 1980)

with the Loch Lomond Readvance ice limit at Menteith. Radiocarbon dates and pollen analyses indicate ages of about 9600 and 8800 years B.P. for the Main and Low Buried Beaches respectively.

A transgression of the sea occurred in early Flandrian times and it formed the Main Post-Glacial Shoreline prior to 6500 years B.P. at or near the maximum extent of the transgression. The deposition of extensive marine and estuarine deposits, called Carse Clays, took place in the estuaries of the Forth and the Tay contemporaneously with the local accumulation of beds of peat. In the Firth of Clyde sand and gravel beaches formed at many places and are especially prominent in Ayrshire between Ayr and Ardrossan, and at Girvan. The sea also connected with the Loch Lomond basin at the maximum of the Flandrian transgression.

Isostatic tilting resulted in the beach sloping gently eastwards in the Forth and Tay areas and to the south and south-west in the Firth of Clyde.

Since the formation of the Main Post-Glacial Shoreline, sea level has fallen and younger, lower shoreline features have formed.

Flandrian

The Flandrian Stage began 10 000 years B.P. and the deposits include the marine and estuarine sediments associated with the Flandrian transgression, river alluvium, peat and blown sand. Movements of sea level are described above.

The climate during the Flandrian improved from arctic conditions at the end of the Devensian to the Climatic Optimum 5000 to 3000 years B.P. and has since become cooler and wetter. The stage is subdivided into five periods, each characterised by a different climatic regime. The subdivision and the corresponding pollen zones are shown in Table 6. The climatic improvement, marking the beginning of Flandrian times, brought widespread periglacial activity to an end and woodland and forest gradually became established. Deforestation, under man's influence, began about 5000 years B.P.

The carse lands of the Forth, Tay and Clyde estuaries are underlain by marine and estuarine deposits laid down during, and subsequent to the Flandrian transgression which reached its maximum prior to 6500 years B.P. The deposits consist of marine clay, silt and rarely sand, with layers of shells. At Flanders Moss, near the Lake of Menteith, peat growth was virtually continuous throughout the period of deposition of the Carse Clays.

The raised beach deposits in the outer parts of the estuaries and in the Firth of Clyde consist of sands and gravels locally containing peat layers.

The growth of hill peat is thought to have been initiated in the Atlantic Period, but basin peat may have started accumulating in the Boreal Period. Large areas of lowland peat have been stripped by man and recent erosion is affecting areas of hill peat.

Alluvium commonly forms thin terraced deposits of silt and sand with lenses of gravel along river banks and in lochs where streams enter.

Areas of blown sand occur at Montrose, near Carnoustie and St Andrews and parts of the East Lothian coast in the east and on the Ayrshire coast between Saltcoats and Prestwick in the west.

Table 6 Subdivision, chronology, etc. of Quaternary sediments in the Midland Valley

STAGE	RADIO-CARBON YEARS B.P.	BLYTT AND SERNANDER PERIODS AND POLLEN ZONES		CLIMATE	RELATIVE SEA-LEVEL	DEPOSITS
FLANDRIAN			VIII-Modern			
	1000	SUB-ATLANTIC	VIII-VIIb	BECOMING COOLER AND WETTER TO PRESENT DAY		
	2000					
	3000	SUB-BOREAL		WARM AND DRY } CLIMATIC OPTIMUM		
	4000					
	5000				NET FALL	HILL PEAT
	6000	ATLANTIC	VIIa	WARM AND WET		
	7000				MAIN POST-GLACIAL SHORELINE	CARSE CLAYS AND PEAT; RAISED BEACH SANDS AND GRAVELS
	8000	BOREAL	V-VI	BECOMING MILDER AND DRIER	FLANDRIAN MARINE TRANSGRESSION MAXIMUM PRIOR TO 6500 BP; NET RISE	
	9000				LOW BURIED RAISED BEACH	
	10000	PRE-BOREAL	IV	COOL (COMPARABLE TO PRESENT DAY); RAPID RISE IN TEMPERATURE	MAIN BURIED RAISED BEACH; NET FALL	
Latter part of LATE DEVENSIAN		YOUNGER DRYAS	III	ARCTIC (LOCH LOMOND READVANCE)	HIGH BURIED RAISED BEACH; NET RISE	LOCH LOMOND READVANCE MORAINE AND OUTWASH
	11000				MAIN LATE-GLACIAL SHORELINE	BURIED GRAVEL LAYER
		ALLERØD	II	LATE-GLACIAL (WINDERMERE) INTERSTADIAL		
	12000					CULFARGIE BEDS; POWGAVIE CLAY
		OLDER DRYAS	Ic	MILDER	NET FALL	CLYDE BEDS
	13000	BØLLING	Ib	(AS WARM AS PRESENT DAY)	MAIN PERTH SHORELINE	
	14000	OLDEST DRYAS	Ia	ARCTIC (GLACIAL RETREAT)	LATE-GLACIAL SHORELINES OF EAST FIFE, ANGUS AND KINCARDINE	ERROL BEDS

16. Economic geology

Fuel and Energy

Coal

Coal remains the single most valuable natural resource in the Midland Valley despite the decline in production since the early part of the century. Maxiumum production took place in 1913 when 42 million tonnes were mined and numerous collieries were in operation. Few collieries are active at the present time and the total annual production from mining and opencast working is currently 11 to 12 million tonnes, mainly of bituminous coal.

The seams still being worked occur in the Limestone Coal Group of Namurian age and in the Lower and Middle Coal Measures of the Westphalian. In addition, there is an important seam in the Upper Limestone Group in the Kincardine Basin which is especially suitable for burning in power stations.

In the early years of the coal-mining industry, attention was concentrated on the more accessible seams at shallow depths. In many cases plans of these workings are not available and possible subsidence, due to the presence of voids at relatively shallow depths, is a major problem in relation to new construction projects.

Large areas of the coalfields are now no longer worked, partly through exhaustion and partly because the remaining coal cannot be worked economically by the mechanised methods currently in use in the industry. Untapped reserves still exist under parts of the Firth of Forth and in many areas there are seams suitable for opencast extraction.

Oil-shales

The oil-shale industry of West Lothian, which flourished from the middle of the last century until its closure in 1962, was initiated by James 'Paraffin' Young. He obtained a patent for the production of 'Paraffin oil' in 1850 and set up a plant using, at first the Boghead Coal from the Lower Coal Measures and then the oil-shales from the Dinantian in a process of destructive distillation to produce oil. The industry reached its maximum productivity in the early years of this century with annual outputs of more than 3.3 million tonnes of oil-shale. The yield of oil from a tonne of shale ranged between 70 and 200 litres.

The remaining resources of oil-shale in the Lothians are fragmented into small pockets by extensive former mining and by geological structure. The quantities of oil which could be recovered from the remaining shale are very small in comparison with today's rate of consumption.

Oil and gas

Small quantities of oil and natural gas have been obtained from Dinantian rocks in the Lothians. The possibility that oil from the oil-shales had migrated and accumulated in sandstone reservoir rocks in the Lower Carboniferous attracted the interest of the oil companies. The main structure of interest has been the D'Arcy–Cousland anticline in Midlothian which has been investigated by drilling. Gas from one well was, at one time, fed into the town gas supply in Musselburgh and another well yielded about 3.5 barrels of oil per day. Both wells have now ceased production. A few other structures, including the Salsburgh Anticline near Airdrie, have also been drilled and small yields of gas obtained from horizons in the Calciferous Sandstone Measures.

Geothermal energy

Preliminary investigations into the possibilities of tapping geothermal energy in the Midland Valley suggest that the western part of the area may be favourable. Rates of heat flow and the geothermal gradient approach minimal economic values locally and if a sufficently permeable aquifer exists at a suitable depth there is the possibility of extracting energy. Experimental trials extracting heat from ground water taken from shallow boreholes have been carried out successfully and they demonstrate the possibility of meeting small-scale local demands for domestic or industrial space heating.

Bulk Minerals

Igneous rock

Igneous rock is quarried in the Midland Valley at a current annual rate of about 9 million tonnes. The rock is used principally for roadstone, concrete aggregate and other constructional uses. The majority of quarries are in basic sills or plugs of Carboniferous age, mainly quartz-dolerite, but Devonian and Carboniferous lavas are also worked and felsites in Lanarkshire give the characteristic reddish brown aggregate.

Sand and gravel

Most sand and gravel produced in the Midland Valley comes from fluvioglacial deposits, but a proportion also comes from raised beach deposits and river alluvium. Sandstone and conglomerate of Carboniferous age are also crushed for high-specification aggregates and to produce sand for moulding or glass-making. Production overall is currently in the range of about 6 to 7 million tonnes per annum.

Resources are scattered throughout the Midland Valley but since transport costs form a considerable proportion of the cost of production, workings tend to be within easy access of the main centres of population.

The gravels consist of a mixture of those rock types present in the area of origin which were sufficiently durable to withstand transport by ice and/or water. Greywacke and lava pebbles are common in the southern part of the area, but pebbles of Highland origin are more abundant in the northern part.

Fireclay

The Carboniferous rocks of the Midland Valley contain some of the most valuable fireclays in the United Kingdom. The main sources are in the Passage Group and in the Lower Coal Measures. The fireclays in the Passage Group of the Central Coalfield have been worked underground and opencast, and those in the Lower Coal Measures have been worked by opencast methods in

1 Blindwells open-cast coal working near Tranent, East Lothian. Stoop and room workings in coal, Limestone Coal Group. (D 3484)

2 Middleton Limestone Mine, Midlothian. Stoop and room workings in Lower Limestone Group. (D 3399)

Plate 16

conjunction with coal.

Most of the current production comes from works on the east side of the Central Coalfield between Linlithgow and Fauldhouse, but there are also workings near Bonnybridge.

Fireclays were also formerly worked in Ayrshire above and below the Passage Group lavas. The Ayrshire Bauxitic Clay overlying the lavas was used mainly for the manufacture of aluminium sulphate.

Large reserves of fireclay have been proved in the Passage Group in the Douglas Coalfield, but they have yet to be exploited.

Bedded iron ores

Clayband and blackband ironstones, which occur mainly in the Limestone Coal Group and the Coal Measures, were the principle source of ore during the rapid growth of the iron and steel industry in Scotland. Ironstone mining reached its maximum in the latter half of last century but declined rapidly in the early part of this century and has now ceased.

Limestone

Limestone in the Midland Valley occurs in relatively thin seams in the Carboniferous and locally in the Upper Devonian. It was worked extensively in the past, both opencast and underground, for agricultural lime, metallurgical flux, stone dust in coal mines and as aggregate and for cement manufacture. Present day production is mainly for agricultural lime and filler powders.

At the present time, the Dockra Limestone in north Ayrshire, the Charlestown Main Limestone in Fife and the North Greens Limestone in Midlothian are being worked.

Large resources adequate for cement manufacture are probably present in only two areas. There are substantial resources of the Dockra and Broadstone limestones in the Beith–Lugton area of north Ayrshire and the Upper Longcraig and Skateraw limestones at East Saltoun in East Lothian.

Many abandoned limestone mines are potentially of use for underground storage or any other purpose for which an underground location would be advantageous. Several mines are of considerable volume extending over several hectares and in some cases have 3 to 4 m of headroom.

Building stone

The use of stone in buildings has declined to a very low level and many of the former sources of supply have been filled in or built over. From time to time, quarries are reopened for supplies of a particular stone for a special requirement or for repair work, but freestone quarrying on a regular basis has virtually ceased.

In the Glasgow area, the best known freestones are the Giffnock and Bishopbriggs Sandstones of the Upper Limestone Group in the Carboniferous and the Ballochmyle Sandstone of the Permian at Mauchline. In Edinburgh most building stones came from the Calciferous Sandstone Measures and include the Binny, Dalmeny, Hermand, Hailes, Dunnet, Granton and Craigleith sandstones. The Craigmillar Sandstone of the Upper Devonian was also used. In the Tay area Lower Devonian Sandstone was used for building and flagstones were formerly quarried in the Forfar district for paving.

Clay and shale

There is very little exploitation of shale and clay in the Midland Valley. Almost 80 per cent of the brick production uses waste material from coal-mining although, in Fife, bricks are made from a mixture of Carboniferous mudstone and boulder clay.

Formerly numerous small brick and tile works thrived using Quaternary clays, till and Carboniferous mudstone.

Peat

Extensive patches of hill peat remain on the higher ground in Ayrshire, on the Campsie and Ochil Hills and to the north-east of Callander. Areas of basin peat occur at Flanders Moss in the upper Forth valley and at several sites in West Lothian, Midlothian and Lanarkshire.

Peat is dug commercially at several sites for horticultural purposes and for use as fuel.

Water supply

Most water supply comes from surface reservoirs at considerable distances from the main areas of demand, but locally groundwater is used for industrial and domestic supply or to augment the public supply.

The aquifers can be broadly subdivided into either rock formations in which groundwater flows mainly through joints or fissures in the rock or unconsolidated granular deposits of Quaternary age in which the groundwater moves through the intergranular voids.

Useful contributions to the public supply and industry are obtained by drilling rock and abstracting water from sedimentary rocks. The Knox Pulpit and Glenvale Formations of the Upper Devonian in Fife give good yields and breweries in Edinburgh obtain water from Upper Devonian and Lower Carboniferous rocks. Elsewhere water is drawn from rocks of Devonian age and from Calciferous Sandstone Measures. The Permian sandstones of the Dumfries area give high yields of groundwater and it is possible that good yields could be obtained from the Mauchline Sandstone in Ayrshire which is of the same age.

Abstraction from unconsolidated deposits is often sufficient for domestic or small industrial needs.

Metalliferous mineral veins

Metalliferous mineralisation is not widespread in the Midland Valley. Baryte, however, has been mined with profit until quite recently from two areas and several small mines and numerous trials were worked in the 19th century and earlier for baryte, copper, lead, iron, silver, nickel and cobalt. The more important occurrences are shown on Figure 42.

Mineral veins are for the most part concentrated in three areas: in Lower Devonian volcanic rocks of the south-western Ochil Hills, in the Lower Carboniferous Clyde Plateau Lavas of the Renfrewshire Hills, and in Silurian, Devonian and basal Carboniferous rocks in the Muirkirk–Lesmahagow area. Outwith these areas mineralisation is mostly associated with Carboniferous volcanic rocks and limestones or with quartz-dolerite intrusions.

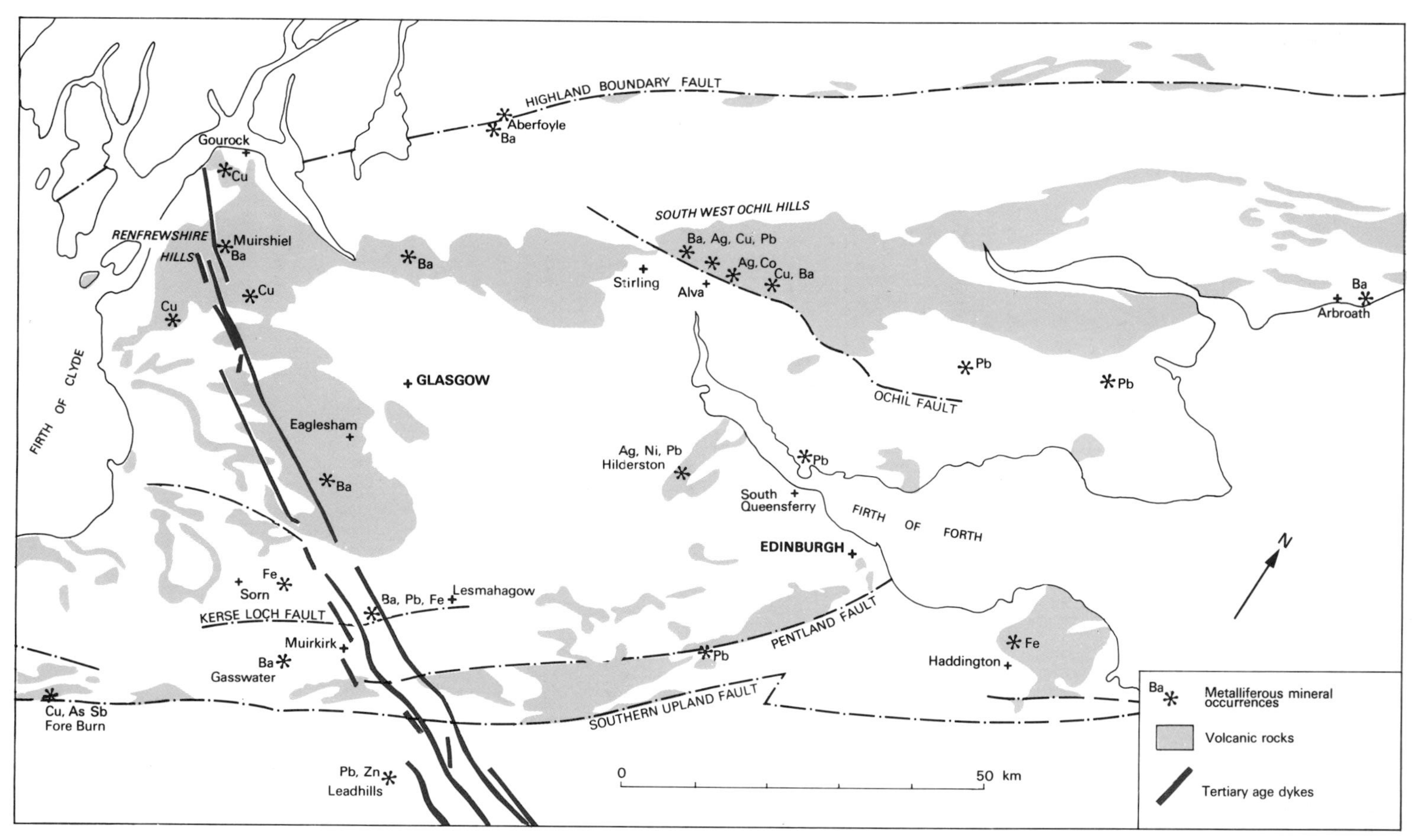

Figure 42 Recorded occurrences of metalliferous veins in the Midland Valley

Baryte

Over 800 000 tonnes of baryte have been produced from the region, accounting for a third of the total UK production between the years 1940 and 1966. Almost all of this output was from vein deposits at Muirshiel in the Renfrewshire Hills (300 000 tonnes; closed 1969) and at Gasswater near Muirkirk (500 000 tonnes; closed 1963).

At the Muirshiel Mine two intersecting NNE–SSW and E–W veins occur at the faulted northern margin of the Misty Law Trachytic Centre, within the Clyde Plateau Lavas. Over forty other recorded baryte veins in the area occur mostly within the trachytes and occupy several fracture directions, but particularly NW–SE fractures which are also occupied by later Tertiary dykes. Baryte veins in trachytic rocks also occur in NW–SE fracture zones south of Eaglesham.

At Gasswater Mine four main veins trend between WNW and NNW in hard sandstone units within an Upper Devonian to Lower Carboniferous sequence. The mine is on the margin of a larger zone of baryte, galena and hematite mineralisation in Silurian to Lower Carboniferous age sediments and Lower Devonian acidic sills around Muirkirk (e.g. Nutberry Hill, River Nethan and Auchinstilloch). Many small veins trend NW–SE, subparallel with the Tertiary dyke swarm which passes through the area, although a major control for vein distribution seems to be the NE–SW Kerse Loch Fault.

In the south-western part of the Ochil Hills, baryte occurs as a gangue mineral in most mineral veins and is the major constituent of the more frequent NW–NNW-trending veins particularly in the areas around Blairlogie and Menstrie. At Aberfoyle worked veins trend between NW and NNW in Lower Devonian sediments near the Highland Boundary Fault.

Copper

In the south-western part of the Ochil Hills many of the veins contain copper, which is particularly notable in the NW-trending baryte veins from which it was formerly extracted at several localities. The principal ores consisted of chalcopyrite, chalcocite and tetrahedrite, with some native copper, chrysocolla and malachite.

Copper mineralisation occurs in a variety of environments in the Renfrewshire Hills. Most occurrences are in the Clyde Plateau Lavas in the form of veins containing chalcocite, bornite, chalcopyrite and secondary malachite in a baryte gangue (e.g. Kaim Mine, Lochwinnoch) but in mines near Gourock disseminations of malachite occur in massive sandstones below the lava sequence. Native copper is recorded from basalts in a few localities (e.g. Boyleston Quarry, Barrhead).

The Fore Burn dioritic complex (p. 46) is affected by widespread tourmalinisation and local disseminated copper-antimony sulphide mineralisation. Traces of cobalt and gold also occur in a breccia zone.

Silver-nickel-cobalt

Small amounts of silver were produced from Hilderston in the Bathgate Hills and from Alva in the Ochil Hills in the early seventeenth and eighteenth century respectively. At Hilderston, vein mineralisation containing baryte and niccolite with some native silver occurred in an E–W fault-plane, adjacent to a quartz-dolerite dyke. The country rocks are clastic sediments of the Lower

Limestone Group within a predominantly volcanic succession. Annabergite (nickel bloom) and erythrite (cobalt bloom) are also recorded. Near Alva, several veins contain native silver and argentite with erythrite in a gangue of calcite and quartz. The most important mine was at Silver Glen, Alva, where both silver and cobalt were extracted.

Lead-zinc

Several veins containing galena and rarely sphalerite cut Silurian sediments in the Nutberry Hill–River Nethan area, Muirkirk. In Fife there are several noted occurrences of galena associated with Carboniferous limestones and/or quartz-dolerite intrusions, and some extraction took place. Galena is present in some of the Ochil mines, including Silver Glen, Alva. At Hilderston, galena and sphalerite were extracted from deeper levels of the silver-bearing vein and from an adjacent vein, both in limestone.

Low grade, stratabound, disseminated zinc-lead mineralisation has recently been discovered in the Lower Limestone Group close to Hilderston Mine and in the Calciferous Sandstone Measures at South Queensferry.

Iron

Hematite has been mined from veins in thick trachyte flows at Garleton, near Haddington and in Silurian sediments at Auchinlongford, near Sorn. Veins of hematite are also recorded from the Auchinstilloch area and ferruginous gouge material occurs in many of the Ochil veins.

Age of the mineralisation

Mineral veins cut rocks ranging in age from Llandovery to topmost Dinantian and are cut by Palaeocene dykes. The mineralisation cannot be regarded as the product of a single event and veins were probably emplaced during several tensional regimes from the late-Caledonian to the Mesozoic.

The rock alteration and mineralisation within the Lower Devonian dioritic complex at Fore Burn may be a late-Caledonian event.

Many of the copper, lead and iron-bearing veins are associated with E–W fracture systems of post-Westphalian age and would seem to be part of the major Hercynian mineralisation episode of the British Isles. Several such veins are closely associated with quartz-dolerite intrusions and a Stephanian age is therefore indicated, in agreement with K-Ar dates on vein gouge clays from the Ochils of 300 to 280 Ma.

Baryte is a constituent of many of the Stephanian age veins but significantly most of the major baryte veins either occupy NW–SE fractures or occur within a 20 km-wide, NW–SE zone, coincident with the main Tertiary dyke swarm (Figure 42). It has been suggested that this north-westerly control may indicate a Tertiary age for the mineralisation. However, it seems more likely that the Tertiary dykes occupy an older mineralised fracture system, possibly dating back to the Carboniferous. Unpublished radiometric dates from the Gasswater baryte deposit of 287 to 270 Ma suggest a late-Carboniferous or Permian age, but the Muirshiel deposit is probably of Triassic age (240–213 Ma). Palaeomagnetic measurements on a baryte-hematite vein from Auchenstilloch point to a still younger Lower to Middle Jurassic age. It is possible that most of the Midland Valley baryte veins originated from an early Mesozoic cycle of magmatic and hydrothermal activity, recognised throughout NW Europe and probably related to the early opening of the North Atlantic.

17 Selected references

The literature on the geology of the region is much too extensive to be given in full. The works listed below, together with their contained bibliographies, will provide a starting point for any reader interested in a particular topic.

General

ANDERTON, R., BRIDGES, P. H., LEEDER, M. R. and SELWOOD, B. W. 1979. *A dynamic stratigraphy of the British Isles.* (London: George Allen and Unwin.)

BLUCK, B. J. (editor). 1973. *Excursion guide to the geology of the Glasgow district.* (Glasgow: Geological Society of Glasgow.)

BOWES, D. R. and LEAKE, B. E. (editors). 1978. Crustal evolution in northwestern Britain and adjacent regions. *Spec. Issue Geol. J.*, No. 10.

CRAIG, G. Y. (editor). 1983. *Geology of Scotland* (2nd edition). (Edinburgh: Scottish Academic Press.)

CRAIG, G. Y. and DUFF, P. McL. D. (editors). 1975. *The geology of the Lothians and south-east Scotland. An excursion guide.* (2nd edition). (Edinburgh: Scottish Academic Press.)

EYLES, V. A., SIMPSON, J. B. and MACGREGOR, A. G. 1949. The Geology of Central Ayrshire. *Mem. Geol. Surv. G.B.*

FORSYTH, I. H. and CHISHOLM, J. I. 1977. The geology of east Fife. *Mem. Geol. Surv. G.B.*

FRANCIS, E. H., FORSYTH, I. H., READ, W. A. and ARMSTRONG, M. 1970. The geology of the Stirling district. *Mem. Geol. Surv. G.B.*

GEORGE, T. N. 1960. The stratigraphical evolution of the Midland Valley. *Trans. Geol. Soc. Glasgow*, Vol. 24, pp. 32–107.

HARRIS, A. L., HOLLAND, C. H. and LEAKE, B. E. (editors). 1979. The Caledonides of the British Isles — reviewed. *Spec. Publ. Geol. Soc. London*, No. 8.

MACGREGOR, A. R. 1973. *Fife and Angus geology. An excursion guide.* (2nd edition). (Edinburgh: Scottish Academic Press.)

MACGREGOR, M., DINHAM, C. H., BAILEY, E. B. and ANDERSON, E. M. 1925. The geology of the Glasgow district. (2nd edition). *Mem. Geol. Surv. G.B.*

MITCHELL, G. H. and MYKURA, W. 1962. The geology of the neighbourhood of Edinburgh. (3rd edition). *Mem. Geol. Surv. G.B.*

RICHEY, J. E., ANDERSON, E. M. and MACGREGOR, A. G. 1930. The geology of north Ayrshire. *Mem. Geol. Surv. G.B.*

SUTHERLAND, D. S. (editor). 1982. *Igneous rocks of the British Isles.* (Chichester: Wiley.)

In addition to the Sheet and District Memoirs of the Geological Survey there are Economic Memoirs describing the geology of the coalfields. References to these memoirs will be found in the corresponding district memoirs.

Introduction (Chapter 1)

BREMNER, A. 1942. The origin of the Scottish river system. *Scot. Geogr. Mag.*, Vol. 58, pp. 15–20, 54–59, 99–203.
GEORGE, T. N. 1960. *q.v.*
— 1965. The geological growth of Scotland. In *The geology of Scotland* (1st edition), pp. 1–48. CRAIG, G. Y. (editor). (Edinburgh: Oliver and Boyd.)
HARLAND, W. B., COX, A. V., LLEWELLYN, P. G., PICKTON, C. A. G., SMITH, A. G. and WALTERS, R. 1982. *A geologic time scale.* (Cambridge: Cambridge University Press.)
LINTON, D. L. 1951. Problems of Scottish scenery. *Scot. Geogr. Mag.*, Vol. 67, pp. 65–85.
MACKINDER, H. J. 1902. *Britain and the British seas.* (London: Heinemann.)
SMITH, A. G., HURLEY, A. M. and BRIDEN, J. C. 1981. *Phanerozoic palaeocontinental world maps.* (Cambridge: Cambridge University Press.)

Pre-Palaeozoic basement (Chapter 2)

BAMFORD, D. 1979. Seismic constraints on the deep geology of the Caledonides of northern Britain. *In* HARRIS, A. L. and OTHERS (editors), *q.v.*
GRAHAM, A. M. and UPTON, B. G. J. 1978. Gneisses in diatremes, Scottish Midland Valley: petrology and tectonic implications. *J. Geol. Soc. London*, Vol. 135, pp. 219–228.
UPTON, B. G. J., ASPEN, P. and CHAPMAN, N. A. 1983. The upper mantle and deep crust beneath the British Isles: evidence from inclusions in volcanic rocks. *J. Geol. Soc. London*, Vol. 140, pp. 105–122.

Ordovician and Silurian (Chapter 3)

LAMONT, A. 1947. Gala-Tarannon beds in the Pentland Hills, near Edinburgh. *Geol. Mag.*, Vol. 84, pp. 193–208 and 289–303.
— 1952. Ecology and correlation of the Pentlandian — a new division of the Silurian System in Scotland. *Rep. 18th Int. Geol. Congr., G.B.*, Part 10, pp. 27–32.
PEACH, B. N. and HORNE, J. 1899. The Silurian rocks of Britain. 1, Scotland. *Mem. Geol. Surv. G.B.*
ROLFE, W. D. I. 1960. The Silurian inlier of Carmichael, Lanarkshire. *Trans. R. Soc. Edinburgh*, Vol. 64, pp. 245–260.
— 1961. The geology of the Hagshaw Hills Silurian inlier, Lanarkshire. *Trans. Edinburgh Geol. Soc.*, Vol. 18, pp. 240–269.
— 1973. Excursion 14. The Hagshaw Hills Silurian inlier. *In* BLUCK, B. J. (editor), *q.v.*
SELDEN, P. A. and WHITE, D. E. 1983. A new Silurian arthropod from Lesmahagow, Scotland. *Spec. Pap. in Palaeontology*, No. 30, pp. 43–49.
TIPPER, J. C. 1976. The stratigraphy of the North Esk inlier, Midlothian. *Scott. J. Geol.*, Vol. 12, pp. 15–22.

Devonian (Chapter 4)

ARMSTRONG, M. and PATERSON, I. B. 1970. The Lower Old Red Sandstone of the Strathmore region. *Rep. Inst. Geol. Sci.*, No. 70/12.
BLUCK, B. J. 1978. Sedimentation in a late orogenic basin: the Old Red Sandstone

of the Midland Valley of Scotland. *In* BOWES, D. R. and LEAKE, B. E. (editors), *q.v.*
— 1980. Evolution of a strike-slip fault-controlled basin, Upper Old Red Sandstone, Scotland. In *Sedimentation in oblique slip mobile zones*, pp. 63–78. READING, H. G. and BALLANCE, P. F. (editors). *Spec. Publ. Int. Assoc. Sedimentol.*, No. 4.
— 1983. Role of the Midland Valley of Scotland in the Caledonian orogeny. *Trans. R. Soc. Edinburgh: Earth Sci.*, Vol. 74, pp. 119–136.
CHISHOLM, J. I. and DEAN, J. M. 1974. The Upper Old Red Sandstone of Fife and Kinross; a fluviatile sequence with evidence of marine incursion. *Scott. J. Geol.*, Vol. 10, pp. 1–30.
DOWNIE, C. and LISTER, T. R. 1969. The Sandy's Creek Beds (Devonian) of Farland Head, Ayrshire. *Scott. J. Geol.*, Vol. 5, pp. 193–206.
HOUSE, M. R., RICHARDSON, J. B., CHALONER, W. G., ALLEN, J. R. L., HOLLAND, C. H. and WESTOLL, T. S. 1977. A correlation of Devonian rocks of the British Isles. *Spec. Rep. Geol. Soc. London*, No. 7, 110 pp.
LUMSDEN, G. I. 1982. Devonian strata in Scotland. *Rep. Inst. Geol. Sci.*, No. 82/1.
MILES, R. S. 1968. The Old Red Sandstone antiarchs of Scotland: Family Bothriolepididae. *Palaeontogr. Soc.* (Monogr.), No. 552, (Vol. 122).
MORTON, D. J. 1979. Palaeogeographical evolution of the Lower Old Red Sandstone basin in the western Midland Valley. *Scott. J. Geol.*, Vol. 15, pp. 97–116.
MYKURA, W. 1960. The Lower Old Red Sandstone igneous rocks of the Pentland Hills. *Bull. Geol. Surv. G.B.*, No. 16, pp. 131–155.
QURESHI, I. R. 1970. A gravity survey in the region of the Highland Boundary Fault in Scotland. *Q.J. Geol. Soc. London*, Vol. 125, pp. 481–502.
READ, W. A. and JOHNSON, S. R. H. 1967. The sedimentology of sandstone formations within the Upper Old Red Sandstone and lowest Calciferous Sandstone Measures west of Stirling. *Scott. J. Geol.*, Vol. 3, pp. 242–267.
RICHARDSON, J. B. 1967. Some British Lower Devonian spore assemblages and their stratigraphical significance. *Rev. Palaeobot. Palynol.*, Vol. 1, pp. 111–129.

Silurian and Devonian igneous activity (Chapter 5)

ARMSTRONG, M. and PATERSON, I. B. 1970. *q.v.*
ELLIOTT, R. B. 1982. The Old Red Sandstone continent: Devonian volcanism. *In* SUTHERLAND, D. S. (editor), *q.v.*
FITTON, J. G., THIRLWALL, M. F. and HUGHES, D. J. 1982. Volcanism in the Caledonian orogenic belt of Britain. In *Andesites*, pp. 611–636. THORPE, R. S. (editor) (Chichester: Wiley.)
FRENCH, W. J., HASSAN, M. D. and WESTCOTT, J. 1979. The petrogenesis of Old Red Sandstone volcanic rocks of the western Ochils, Stirlingshire. *In* HARRIS, A. L and OTHERS (editors), *q.v.*
GANDY, M. K. 1975. The petrology of the Lower Old Red Sandstone lavas of the eastern Sidlaw Hills, Perthshire, Scotland. *J. Petrol.*, Vol. 16, pp. 189–211.
KOKELAAR, B. P. 1982. Fluidization of wet sediments during the emplacement and cooling of various igneous bodies. *J. Geol. Soc. London*, Vol. 139, pp. 21–34.
MYKURA, W. 1960. *q.v.*
PATERSON, I. B. and HARRIS, A. L. 1969. Lower Old Red Sandstone ignimbrites from Dunkeld, Perthshire. *Rep. Inst. Geol. Sci.*, No. 69/7.
STILLMAN, C. J. and FRANCIS, E. H. 1979. Caledonian volcanism in Britain and Ireland. *In* HARRIS, A. L. and OTHERS (editors), *q.v.*
THIRLWALL, M. F. 1981. Implications for Caledonian plate tectonic models of

chemical data from volcanic rocks of the British Old Red Sandstone. *J. Geol. Soc. London*, Vol. 138, pp. 123–138.
— 1982. Systematic variation in chemistry and Nd-Sr isotopes across a Caledonian calc-alkaline volcanic arc: implications for source materials. *Earth Planet. Sci. Lett.*, Vol. 58, pp. 27–50.
— 1983a. Discussion on implications for Caledonian plate tectonic models of chemical data from volcanic rocks of the British Old Red Sandstone. *J. Geol. Soc. London*, Vol. 140, pp. 315–318.
— 1983b. Isotope geochemistry and origin of calc-alkaline lavas from a Caledonian continental margin volcanic arc. *J. Volcanol. Geotherm. Res.*, Vol. 18, pp. 589–631.

Carboniferous (Chapters 6 to 10)

BELT, E. S., FRESHNEY, E. C. and READ, W. A. 1967. Sedimentology of Carboniferous cementstone facies, British Isles and eastern Canada. *J. Geol.*, Vol. 75, pp. 711–721.
BRAND, P. J. 1977. The fauna and distribution of the Queenslie Marine Band (Westphalian) in Scotland. *Rep. Inst. Geol. Sci.*, No. 77/18.
— 1983. Stratigraphical palaeontology of the Westphalian of the Ayrshire Coalfield, Scotland. *Trans. R. Soc. Edinburgh: Earth Sci.*, Vol. 73, pp. 173–190.
— ARMSTRONG, M. and WILSON, R. B. 1980. The Carboniferous strata at the Westfield Opencast Site, Fife, Scotland. *Rep. Inst. Geol. Sci.*, No. 79/11.
BOTT, M. H. P. and JOHNSON, G. A. L. 1967. The controlling mechanism of Carboniferous cyclic sedimentation. *Q.J. Geol. Soc. London*, Vol. 122, pp. 421–441.
BROWNE, M. A. E. 1980. Stratigraphy of the lower Calciferous Sandstone Measures in Fife. *Scott. J. Geol.*, Vol. 16, pp. 321–328.
CURRIE, E. D. 1954. Scottish Carboniferous goniatites. *Trans. R. Soc. Edinburgh,* Vol. 62, pp. 527–602.
DAVIES, A. 1972. Carboniferous rocks of the Muirkirk, Gass Water and Glenmuir areas of Ayrshire. *Bull. Geol. Surv. G.B.*, No. 40, pp. 1–49.
— 1974. The Lower Carboniferous (Dinantian) sequence at Spilmersford, East Lothian, Scotland. *Bull. Geol. Surv. G.B.*, No. 45, pp. 1–38.
FORSYTH, I. H. 1978. The lower part of the Limestone Coal Group in the Glasgow district. *Rep. Inst. Geol. Sci.*, No. 78/29.
— 1979. The Lower Coal Measures of central Glasgow. *Rep. Inst. Geol. Sci.*, No. 79/4.
— 1982. The stratigraphy of the Upper Limestone Group ($E_1$and E_2 stages of the Namurian) in the Glasgow district. *Rep. Inst. Geol. Sci.*, No. 82/4.
GEORGE, T. N. 1978. Eustasy and tectonics: sedimentary rhythms and stratigraphical units in British Dinantian correlation. *Proc. Yorkshire Geol. Soc.*, Vol. 42, pp. 229–254.
— JOHNSON, G. A. L., MITCHELL, M., PRENTICE, J. E., RAMSBOTTOM, W. H. C., SEVASTOPULO, G. D. and WILSON, R. B. 1976. A correlation of Dinantian rocks in the British Isles. *Spec. Rep. Geol. Soc. London*, No. 7.
GOODLET, G. A. 1957. Lithological variation in the Lower Limestone Group of the Midland Valley of Scotland. *Bull. Geol. Surv. G.B.*, No. 12, pp. 52–65.
GREENSMITH, J. T. 1965. Calciferous Sandstone Series sedimentation at the eastern end of the Midland Valley of Scotland. *J. Sediment. Petrol.*, Vol. 35, pp. 223–242.
HILL, D. 1938–41. Carboniferous rugose corals of Scotland. *Palaeontogr. Soc.* (Monogr.)
LUMSDEN, G. I. 1964. The Limestone Coal Group of the Douglas Coalfield,

Lanarkshire. *Bull. Geol. Surv. G.B.*, No. 21, pp. 37–71.
— 1965. The base of the Coal Measures in the Douglas Coalfield, Lanarkshire. *Bull. Geol. Surv. G.B.*, No. 22, pp. 80–91.
— 1967. The Carboniferous Limestone Series of Douglas, Lanarkshire. *Bull. Geol. Surv. G.B.*, No. 26, pp. 1–22.
— 1967. The Upper Limestone Group and Passage Group of Douglas, Lanarkshire. *Bull. Geol. Surv. G.B.*, No. 27, pp. 17–48.
— and CALVER, M. A. 1958. The stratigraphy and palaeontology of the Coal Measures of the Douglas Coalfield. *Bull. Geol. Surv. G.B.*, No. 15, pp. 32–70.
— and WILSON, R. B. 1978. Stratigraphical classification of the Carboniferous succession of central Scotland. *C.R. 8e Congr. Int. Stratigr. Geol. Carbonif.*, Vol. 2, pp. 27–36.
MCLEAN, A. C. and DEEGAN, C. E. 1976. A synthesis of the solid geology of the Firth of Clyde region *In* MCLEAN, A. C. and DEEGAN, C. E. (editors). The solid geology of the Clyde sheet (55°N/6°W). *Rep. Inst. Geol. Sci.*, No. 78/9.
MONRO, S. K., LOUGHNAN, F. C. and WALKER, M. C. 1983. The Ayrshire bauxitic clay: an allochthonous deposit? In *Residual Deposits*, pp. 47–58. WILSON, R. C. L. (editor). *Spec. Publ. Geol. Soc. London*, No. 11.
MUIR, R. O. 1963. Petrography and provenance of the Millstone Grit of central Scotland. *Trans. Edinburgh Geol. Soc.*, Vol. 19, pp. 439–485.
MYKURA, W. 1960. The replacement of coal by limestone and the reddening of Coal Measures in the Ayrshire Coalfield. *Bull. Geol. Surv. G.B.*, No. 16, pp. 69–109.
— 1967. The Upper Carboniferous rocks of south-west Ayrshire. *Bull. Geol. Surv. G.B.*, No. 26, pp. 23–48.
NEVES, R., READ, W. A. and WILSON, R. B. 1965. Note on recent spore and goniatite evidence from the Passage Group of the Scottish Upper Carboniferous succession. *Scott. J. Geol.*, Vol. 1, pp. 185–188.
— GUEINN, K. J., CLAYTON, G., IOANNIDES, N. S., NEVILLE, R. S. W. and KRUSZEWSKA, K. 1973. Palynological correlations within the Lower Carboniferous of Scotland and northern England. *Trans. R. Soc. Edinburgh*, Vol. 69, pp. 23–70.
RAMSBOTTOM, W. H. C. 1977. Major cycles of transgression and regression (Mesothems) in the Namurian. *Proc. Yorkshire Geol. Soc.*, Vol. 41, pp. 261–291.
— 1977. Correlation of the Scottish Upper Limestone Group (Namurian) with that of the North of England. *Scott. J. Geol.*, Vol. 13, pp. 327–330.
— CALVER, M. A., EAGAR, R. M. C., HODSON, F., HOLLIDAY, D. W., STUBBLEFIELD, C. J. and WILSON, R. B. 1978. A correlation of Silesian rocks in the British Isles. *Spec. Rep. Geol. Soc. London*, No. 10.
READ, W. A. and DEAN, J. M. 1967. A quantitative study of a sequence of coal-bearing cycles in the Namurian of central Scotland, 1. *Sedimentology*, Vol. 9, pp. 137–156.
— and COLE, A. J. 1971. Some Namurian (E_2) paralic sediments in central Scotland: an investigation of depositional environment and facies changes using iterative-fit trend-surface analysis. *J. Geol. Soc. London*, Vol. 127, pp. 137–176.
SMITH, A. G., HURLEY, A. M. and BRIDEN, J. C. 1981. *q.v.*
THOMSON, M. E. 1978. IGS Studies of the geology of the Firth of Forth and its approaches. *Rep. Inst. Geol. Sci.*, No. 77/17.
WEIR, J. and LEITCH, D. 1936. The zonal distribution of the non-marine lamellibranchs in the Coal Measures of Scotland. *Trans. R. Soc. Edinburgh*, Vol. 57, pp. 697–751.
WILSON, R. B. 1966. A study of the Neilson Shell Bed, a Scottish Lower Carboniferous marine shale. *Bull. Geol. Surv. G.B.*, No. 24, pp. 105–128.

— 1967. A study of some Namurian marine faunas of central Scotland. *Trans. R. Soc. Edinburgh*, Vol. 66, pp. 445–490.
— 1974. A study of the Dinantian marine faunas of south-east Scotland. *Bull. Geol. Surv. G.B.*, No. 46, pp. 35–65.
— 1979. The base of the Lower Limestone Group (Viséan) in North Ayrshire. *Scott. J. Geol.*, Vol. 15, pp. 313–319.
— 1983. Note on the correlation of the Upper Limestone Group (Namurian) in south Ayrshire. *Scott. J. Geol.*, Vol. 19, pp. 183–188.

Permian and Triassic (Chapter 11)

McLean, A. C. and Deegan, C. E. 1978. *q.v.*
Mykura, W. 1965. The age of the lower part of the New Red Sandstone of south-west Scotland. *Scott. J. Geol.*, Vol. 1, pp. 9–18.
Smith, A. G., Hurley, A. M. and Briden, J. C. 1981. *q.v.*
Smith, D. B., Brunstrom, R. G. W., Manning, P. I., Simpson, S. and Shotton, F. W. 1974. A correlation of Permian rocks in the British Isles. *Spec. Rep. Geol. Soc. London*, No. 5.
Thomson, M. E. 1978. *q.v.*
Wagner, R. H. 1983. A Lower Rotliegend flora from Ayrshire. *Scott. J. Geol.*, Vol. 19, pp. 135–155.
Warrington, G., Audley-Charles, M. G., Elliott, R. E., Evans, W. B., Ivimey-Cook, H. C., Kent, P. E., Robinson, P. L., Shotton, F. W. and Taylor, F. M. 1980. A correlation of Triassic rocks in the British Isles. *Spec. Rep. Geol. Soc. London*, No. 13, 77 pp.

Carboniferous and Permian igneous activity (Chapter 12)

Black, G. P. 1966. *Arthur's Seat. A history of Edinburgh's volcano.* (Edinburgh: Oliver and Boyd.)
Chapman, N. A. 1976. Inclusions and megacrysts from undersaturated tuffs and basanites, east Fife, Scotland. *J. Petrol.*, Vol. 17, pp. 472–498.
Craig, P. M. and Hall, I. H. S. 1975. The Lower Carboniferous rocks of the Campsie–Kilpatrick area. *Scott. J. Geol.*, Vol. 11, pp. 171–174.
De Souza, H. A. F. 1974. Potassium-argon ages of Carboniferous igneous rocks from East Lothian and the south of Scotland. Unpublished M.Sc. thesis, University of Leeds.
— 1979. The geochronology of Scottish Carboniferous volcanism. Unpublished Ph.D. thesis, University of Edinburgh.
Fitch, F. J., Miller, J. A. and Williams, S. C. 1970. Isotopic ages of British Carboniferous rocks. *C.R. 6e Congr. Int. Stratigr. Geol. Carbonif. (Sheffield, 1967)*, Vol. 2, pp. 771–789.
Forsyth, I. H. and Rundle, C. C. 1978. The age of the volcanic and hypabyssal rocks of east Fife. *Bull. Geol. Surv. G.B.*, No. 60, pp. 23–29.
Francis, E. H. 1978a. Igneous activity in a fractured craton: Carboniferous volcanism in northern Britain. *In* Bowes, D. R. and Leake, B. E. (editors), *q.v.*
— 1978b. The Midland Valley as a rift, seen in connection with the late Palaeozoic European Rift System. In *Tectonics and Geophysics of Continental Rifts*, pp. 133–147. Ramberg, I. B. and Neumann, E.-R. (editors). (Dordrecht: Reidel.)
— 1982. Magma and sediment - 1. Emplacement mechanism of late Carboniferous tholeiite sills in northern Britain. *J. Geol. Soc. London*, Vol. 139, pp. 1–20.
Hall, J. 1974. A seismic reflection survey of the Clyde Plateau lavas in north Ayrshire and Renfrewshire. *Scott. J. Geol.*, Vol. 9, pp. 253–279.

JOHNSTONE, G. S. 1965. The volcanic rocks of the Misty Law–Knockside Hills district, Renfrewshire. *Bull. Geol. Surv. G.B.*, No. 22, pp. 53–64.
MACDONALD, J. G. and WHYTE, F. 1981. Petrochemical evidence for the genesis of a Lower Carboniferous transitional basaltic suite in the Midland Valley of Scotland. *Trans. R. Soc. Edinburgh: Earth Sci.*, Vol. 72, pp. 75–88.
MACDONALD, R. 1975. Petrochemistry of the early Carboniferous (Dinantian) lavas of Scotland. *Scott. J. Geol.*, Vol. 11, pp. 269–314.
— 1980. Trace element evidence for mantle heterogeneity beneath the Scottish Midland Valley in the Carboniferous and Permian. *Philos. Trans. R. Soc. London*, Ser. A, Vol. 297, pp. 245–257.
— GOTTFRIED, D., FARRINGTON, M. J., BROWN, F. W. and SKINNER, N. G. 1981. The geochemistry of a continental tholeiitic suite: late Palaeozoic quartz-dolerite dykes of Scotland. *Trans. R. Soc. Edinburgh: Earth Sci.*, Vol. 72, pp. 57–74.
— THOMAS, J. E. and RIZZELLO, S. A. 1977. Variations in basalt chemistry with time in the Midland Valley province during the Carboniferous and Permian. *Scott. J. Geol.*, Vol. 13, pp. 11–22.
MACGREGOR, A. G. 1928. The classification of Scottish Carboniferous olivine-basalts and mugearites. *Trans. Geol. Soc. Glasgow*, Vol. 18, pp. 324–360.
MACINTYRE, R. M., CLIFF, R. A. and CHAPMAN, N. A. 1981. Geochronological evidence for phased volcanic activity in Fife and Caithness necks, Scotland. *Trans. R. Soc. Edinburgh: Earth Sci.*, Vol. 72, pp. 1–7.
MONRO, S. K., LOUGHNAN, F. C. and WALKER, M. C. 1983. *q.v.*
MYKURA, W. 1965. White trap in some Ayrshire coals. *Scott. J. Geol.*, Vol. 1, pp. 176–184.
— 1967. *q.v.*
UPTON, B. G. J. (editor). 1969. *Field excursion guide to the Carboniferous volcanic rocks of the Midland Valley of Scotland.* (Edinburgh: Edinburgh Geological Society.)
— 1982. Carboniferous to Permian volcanism in the stable foreland. *In* SUTHERLAND, D. S. (editor), *q.v.*
— ASPEN, P. and CHAPMAN, N. A. 1983. *q.v.*
WALKER, F. 1935. The late Palaeozoic quartz-dolerites and tholeiites of Scotland. *Mineral. Mag.*, Vol. 24, pp. 131–159.
WHYTE, F. 1968. Lower Carboniferous volcanic vents in the west of Scotland. *Bull. Volcanol.*, Vol. 32, pp. 253–268.
— and MACDONALD, J. G. 1974. Lower Carboniferous vulcanicity in the northern part of the Clyde Plateau. *Scott. J. Geol.*, Vol. 10, pp. 187–198.

Tertiary igneous intrusions (Chapter 13)

MACINTYRE, R. M., MCMENAMIN, T. and PRESTON, J. 1975. K-Ar results from western Ireland and their bearing on the timing and siting of Thulean magmatism. *Scott. J. Geol.*, Vol. 11, pp. 227–249.
MYKURA, W. 1967. *q.v.*
SPEIGHT, J. M., SKELHORN, R. R., SLOAN, T. and KNAPP, R. J. 1982. The dyke swarms of Scotland. *In* SUTHERLAND, D. S. (editor), *q.v.*
THOMPSON, R. N. 1982a. Geochemistry and magma genesis. *In* SUTHERLAND, D. S. (editor), *q.v.*
— 1982b. Magmatism of the British Tertiary Volcanic Province. *Scott. J. Geol.*, Vol. 18, pp. 49–107.

Structure (Chapter 14)

ALLAN, D. A. 1928. The geology of the Highland Border from Tayside to Noranside. *Trans. R. Soc. Edinburgh*, Vol. 56, pp. 57–88.

— 1940. The geology of the Highland Border from Glen Almond to Glen Artney. *Trans. R. Soc. Edinburgh*, Vol. 60, pp. 171–193.
ANDERSON, E. M. 1951. *The dynamics of faulting and dyke formation with application to Britain*. (2nd edition). (Edinburgh: Oliver and Boyd.)
ANDERSON, J. G. C. 1947. The geology of the Highland Border, Stonehaven to Arran. *Trans. R. Soc. Edinburgh*, Vol. 61, pp. 479–515.
BAMFORD, D. 1979. *q.v.*
BLUCK, B. J. 1978. *q.v.*
— 1980. *q.v.*
— 1983. *q.v.*
DEWEY, J. F. 1971. A model for the Lower Palaeozoic evolution of the southern margin of the early Caledonides of Scotland and Ireland. *Scott. J. Geol.*, Vol. 7, pp. 219–240.
GEORGE, T. N. 1960. *q.v.*
HALL, J. 1974. *q.v.*
KENNEDY, W. Q. 1958. Tectonic evolution of the Midland Valley of Scotland. *Trans. Geol. Soc. Glasgow*, Vol. 23, pp. 107–133.
LAMBERT, R. ST. J. and MCKERROW, W. S. 1976. The Grampian orogeny. *Scott. J. Geol.*, Vol. 12, pp. 271–292.
LEEDER, M. R. 1982. Upper Palaeozoic basins of the British Isles — Caledonide inheritance versus Hercynian plate margin processes. *J. Geol. Soc. London*, Vol. 139, pp. 479–491.
MCKERROW, W. S., LEGGETT, J. K. and EALES, M. H. 1977. Imbricate thrust model of the Southern Uplands of Scotland. *Nature, London*, Vol. 267, 237–239.
MCLEAN, A. C. 1966. A gravity survey in Ayrshire and its geological interpretation. *Trans. R. Soc. Edinburgh*, Vol. 66, pp. 239–265.
— and DEEGAN, C. E. 1976. *q.v.*
PATERSON, E. M. 1954. Notes on the tectonics of the Greenock–Largs Uplands and the Cumbraes. *Trans. Geol. Soc. Glasgow*, Vol. 21, pp. 430–435.
PHILLIPS, W. E. A., STILLMAN, C. J. and MURPHY, T. 1976. A Caledonian plate tectonic model. *J. Geol. Soc. London*, Vol. 132, pp. 579–609.
RUSSELL, M. J. and SMYTH, D. K. 1983. Origin of the Oslo Graben in relation to the Hercynian–Alleghenian Orogeny and lithospheric rifting in the North. Atlantic. *Tectonophysics*, Vol. 94, pp. 457–472.
THOMSON, M. E. 1978. *q.v.*

Quaternary (Chapter 15)

BISHOP, W. W. and COOPE, G. R. 1979. Stratigraphical and faunal evidence for late-Glacial and early Flandrian environments in south-west Scotland. In *Studies in the Scottish late-Glacial environment*, GRAY, J. M. and LOWE, J. J. (editors). (Oxford: Pergamon.)
BROWNE, M. A. E., MCMILLAN, A. M. and GRAHAM, D. K. 1983. A late-Devensian marine and non-marine sequence near Dumbarton, Strathclyde. *Scott. J. Geol.*, Vol. 19, pp. 229–234.
CULLINGFORD, R. A. and SMITH, D. E. 1966. Late-glacial shorelines in eastern Fife. *Trans. Inst. Br. Geogr.*, Vol. 39, pp. 31–51.
— — 1980. Late Devensian raised shorelines in Angus and Kincardineshire, Scotland. *Boreas*, Vol. 9, pp. 21–38.
DAWSON, A. G. 1980. The low rock platform in western Scotland. *Proc. Geol. Assoc.*, Vol. 91, pp. 339–344.
FORSYTH, I. H. and CHISHOLM, J. I. 1977. *q.v.*
GOODLET, G. A. 1964. The kamiform deposits near Carstairs, Lanarkshire. *Bull. Geol. Surv. G.B.*, No. 21, pp. 175–196.

GRAY, J. M. 1978. Low-level shore platforms in the south-west Scottish Highlands: altitude, age and correlation. *Trans. Inst. Br. Geogr.*, Vol. 3, pp. 151–164.
KIRKBY, R. P. 1968. The ground moraines of Midlothian and East Lothian. *Scott. J. Geol.*, Vol. 4, pp. 209–220.
MCLELLAN, A. G. 1969. The last glaciation and deglaciation of central Lanarkshire. *Scott. J. Geol.*, Vol. 5, pp. 248–268.
MENZIES, J. 1981. Investigations into the Quaternary deposits and bedrock topography of central Glasgow. *Scott. J. Geol.*, Vol. 17, pp. 155–168.
MITCHELL, G. F., PENNY, L. F., SHOTTON, F. W. and WEST, R. G. 1973. A correlation of Quaternary deposits in the British Isles. *Spec. Rep. Geol. Soc. London*, No. 4, 99 pp.
PATERSON, I. B., ARMSTRONG, M. and BROWNE, M. A. E. 1981. Quaternary estuarine deposits in the Tay–Earn area, Scotland. *Rep. Inst. Geol. Sci.*, No. 81/7.
PEACOCK, J. D., GRAHAM, D. K. and WILKINSON, I. P. 1978. Late-Glacial and post-Glacial marine environments at Ardyne, Scotland, and their significance in the interpretation of the history of the Clyde sea area. *Rep. Inst. Geol. Sci.*, No. 78/17.
PEARS, N. 1975. The growth rate of hill peats in Scotland. *Geol. Foren. Stockholm Forh.*, Vol. 97, pp. 265–270.
ROLFE, W. D. I. 1966. Woolly rhinoceros from the Scottish Pleistocene. *Scott. J. Geol.*, Vol. 2, pp. 252–258.
SISSONS, J. B. 1974. The Quaternary in Scotland: a review. *Scott. J. Geol.*, Vol. 10, pp. 311–337.
— 1976. *The geomorphology of the British Isles: Scotland.* (London: Methuen.)
— 1981. The last Scottish ice-sheet: facts and speculative discussion. *Boreas*, Vol. 10, pp. 1–17.
THOMSON, M. E. and EDEN, R. A. 1977. Quaternary deposits of the central North Sea, 3. The Quaternary sequence in the west-central North Sea. *Rep. Inst. Geol. Sci.*, No. 77/12.

Economic geology (Chapter 16)

ALLEN, P. M., COOPER, D. C., PARKER, M. E., EASTERBROOK, G. D. and HASLAM, H. W. 1982. Mineral exploration in the area of the Fore Burn igneous complex, south-western Scotland. *Mineral Reconnaissance Programme, Rep. Inst. Geol. Sci.*, No. 55.
CADELL, H. M. 1925. The Hilderston silver mine. In *The rocks of West Lothian*, pp. 359–378. (Edinburgh: Oliver and Boyd.)
CARRUTHERS, R. G., CALDWELL, W., BAILEY, E. M. and CONACHER, H. R. J. 1927. The oil-shales of the Lothians (3rd edition). *Mem. Geol. Surv. G.B.*
EVANS, A. M. and EL-NIKHELY, A. 1982. Palaeomagnetic age for mineralisation at Auchenstilloch, Lanarkshire, Scotland. *Trans. Instn. Min. Metall., Sect. B: Appl. Earth Sci.*, Vol. 91, pp. 43–44.
FALCON, N. L. and KENT, P. E. 1960. Geological results of petroleum exploration in Britain 1945–1957. *Mem. Geol. Soc. London*, No. 2
HALL, I. H. S., GALLAGHER, M. J. and OTHERS. 1982. Investigation of polymetallic mineralisation in Lower Devonian volcanics near Alva, central Scotland. *Mineral Reconnaissance Programme Rep. Inst. Geol. Sci.*, No. 53.
HOBSON, G. V. 1959. Barytes in Scotland with special reference to Gasswater and Muirshiels Mines. In *The future of non-ferrous mining in Gt. Britain and Ireland*, pp. 85–100. (London: Instn. Min. Metall.)
INESON, P. R. and MITCHELL, J. G. 1974. K-Ar isotopic age determinations from

some Scottish mineral localities. *Trans. Instn. Min. Metall., Sect. B: Appl. Earth Sci.*, Vol. 83, 13–18.
MACGREGOR, A. G. 1944. Barytes in Central Scotland. *Wartime Pamphlet Geol. Surv. G.B.: Scotland*, No. 38.
MACGREGOR, M., LEE, G. W. and WILSON, G. V. 1920. The iron ores of Scotland *Spec. Rep. Miner. Resour. Mem. Geol. Surv. G.B.*, No. 11.
MOORE, D. J. 1979. The baryte deposits of central and southern Scotland. Unpublished Ph.D thesis, University of Leeds.
PARNELL, J. 1983. Stromatolite-hosted mineralisation in the Oil-Shale Group, Scotland. *Trans. Instn. Min. Metall., Sect. B: Appl. Earth Sci.*, Vol. 92, pp. 98–99.
ROBERTSON, T., SIMPSON, J. B. and ANDERSON, J. G. C. 1949. The Limestones of Scotland. *Mem. Geol. Surv. G.B.*
SCOTT, B. 1967. Barytes mineralisation at Gasswater mine, Ayrshire, Scotland. *Trans. Instn. Min. Metall., Sect. B: Appl. Earth Sci.*, Vol. 76, pp. 40–51.
STEPHENSON, D. 1983. Polymetallic mineralisation in Carboniferous rocks at Hilderston, near Bathgate, central Scotland. *Mineral Reconnaissance Programme Rep. Inst. Geol. Sci.*, No. 68.
— and COATS, J. S. 1983. Baryte and copper mineralisation in the Renfrewshire Hills, central Scotland. *Mineral Reconnaissance Programme Rep. Inst. Geol. Sci.*, No. 67.
WILSON, G. V. 1921. The Pb, Zn, Cu and Ni ores of Scotland. *Spec. Rep. Miner. Resour. Mem. Geol. Surv. G.B.*, No. 17.
— 1922. The Ayrshire Bauxitic Clay. *Mem. Geol. Surv. G.B.*

Index

Printed in the UK for HMSO
Dd 737383 C200